PSYCHE and SYNAPSE
EXPANDING WORLDS

The Role of Neurobiology
in Emotions, Behavior,
Thinking, and Addiction
for Non-Scientists
Second Edition

Harriette C Johnson

ISBN 0-9759095-0-9

Design:
    Eileen Claveloux

Cover Design:
    Eileen Claveloux

Graphics:
    Diana Hardina,Upstream Advertising Graphics,
        Shelburne Falls, Massachusetts
    Timothy Paulson

Prepress Director:
    Jeff Harris

Printed in the United States

To Paul, my treasure

ACKNOWLEDGMENTS

Acknowledgments are made to the following for reproduction of brain images: Alan Zametkin, Clinical Brain Imaging, Daniel Weinberger, Clinical Brain Disorders Branch, John Hsiao, Experimental Therapeutics Branch, and Mark George, Biological Psychiatry Branch, National Institute of Mental Health, National Institutes of Health, US Department of Health and Human Services, Rockville, Maryland; National Alliance for the Mentally Ill, Campaign To End Discrimination, Arlington, Virginia; Lewis Baxter, University of Alabama; Harry Chugani, Wayne State University School of Medicine, Detroit, Michigan, and the Academic Press for permission to use a graph of developmental patterns in brain glucose utilization; the *New England Journal of Medicine* for the use of Dr. Zametkin's PET scans; and Yale University Press for permission to include an adaptation of three brain sketches from *Wider than the Sky*.

Steven Hyman, provost, Harvard University and director of the National Institute of Mental Health, 1996-2001, graciously gave permission to incorporate his material on substance abuse and addiction. Many thanks to Paul Sirovatka, science writer, Office of Science Policy and Program Planning, National Institute of Mental Health, for his support and creative efforts.

I want to thank Gerald Edelman, founder and director of the Neurosciences Institute in La Jolla, California, for his generous consultation, patient explanations, and dazzling metaphors that helped me come closer to understanding the theory of neuronal group selection.

My thanks to Eileen Gambrill of the University of California, Berkeley, for her editorial leadership in introducing complex adaptive systems to social work educators, and to Leonard Gibbs of the University of Wisconsin-Eau Claire for giving us tools to implement evidence-based social work practice. Their work enriches this second edition of *Psyche and Synapse.*

Much appreciation goes to Ludwig Geismar, professor emeritus of Rutgers University, for three decades of supportive commentary sprinkled with whimsical humor.

Marcia Brubeck, Joseph D'Amico, and Naomi Pines Gitterman helped me expand the book's connection to practice. Professor emerita Joan Laird of the Smith College School for Social Work deserves credit for conceiving this project, as does professor and former dean Anita Lightburn for supporting it. Kay Davidson, dean of the University of Connecticut School of Social Work has supported my work in countless ways. Catherine Havens, associate dean, has been unfailingly helpful in finding solutions to knotty problems.

I'm very grateful to Michie Hesselbrock, director of the doctoral program at the University of Connecticut School of Social Work, for providing opportunities for me to expand my knowledge of substance abuse and addiction. David Cournoyer, anthropologist, scholar of child development, and also an associate dean at the School of Social Work, responded enthusiastically to the first edition and spurred me on to write the second with his three-word injunction: "I want more."

Amy Brenner-Leonard, Marcia Brubeck, and Deborah McCutcheon have helped with incisive critiques and stylistic suggestions. Amy has risen to the occasion repeatedly—whatever the occasion happens to be. My research assistants Jill Haga and Jack Lu and my husband Paul Brenner have helped greatly by reading parts of the manuscript and offering suggestions. Jill Haga researched and drafted Part VI on multiple routes to quality of life, contributing research competence enriched by her expertise in the subject area. Ashton Haga (age 14) has done an admirable job helping with computer issues, and thanks also go to Michaela Haga (age 11) for her help with clerical matters.

Jennifer Moore Fisk of Deerfield Valley Publishing has been a major source of support, inspiration, and laughter, not to mention tolerance of my foibles. Jeff Harris has masterfully negotiated challenging computer prepress mazes. Jamie Thaman has provided insightful editorial assistance as well as flexibility in a time crunch. My dear friend Helen Sootin Smith (1936-2003), who edited the first edition of the book, has left an indelible imprint on all facets of my work.

My teachers, Paul Lindale and Eileen Claveloux of Greenfield Community College and Neil Carlson of the University of Massachusetts, Amherst, have inspired and enlightened me. Colleagues at the University of Connecticut Elizabeth D'Amico, Cheryl Stockford, Susan Hamilton Peck, Michael Genovese, Candace McCann, and Linda Benoit have provided helpful suggestions and faithful teaching of neuroscience content to social work students. Special appreciation goes to Joseph D'Amico and Jill Haga for their contributions of clinical material, and to anonymous donors of case material who renounced mention in order to protect their clients.

Eileen Claveloux has added artistic flair, design know-how, and unfailing composure, not to mention generosity in giving up days of vacation time to meet a burdensome deadline. Diana Hardina deserves enormous credit for her clever graphics and her unique humor.

Much appreciation is due to Tom Boisvert, chair of the Art Department, and Greenfield Community College, an outstanding community resource in Western Massachusetts, for making it possible for me to develop animated graphics that bring to life some of the neurobiological events described in the book. I'm grateful to Tim Paulson for getting me started with animations, and for helping me out in a crunch with some final touches. Many thanks to Peter Blanchette, Archguitar Music, for helping me record the title music and voice-overs that accompany neuroscience animations on a CD-ROM, available on request. Emily Estes (age 15) played drums with pizzazz, and Paul Brenner added his impeccable beat with "all natural" percussion—on the bones!

Jean Burch Harrison-Siegler, neuroscientist, naturalist, activist, and lifelong friend, added one more gift to the many she's given me, a critical review of my material on the hypothalamic-pituitary-adrenal axis. Before that, she led me across a magical lake in northern New England to see a family of loons, a species known to endorse equality in parental roles. Their chick rode confidently on the back of one of its parents, but I will never know which parent it was.

Two amazing people deserve special mention. Jan Lambert, University of Connecticut social work librarian, has helped me often and tirelessly for the past 15 years, locating resources, instructing me and my students in the use of all manner of electronic tools, resolving computer glitches, ordering books and videos, obtaining elusive materials, and performing countless other tasks, always with unfailing patience and kindness. Virginia Starkie, my secretary for 17 years, has been—well—indispensable. I couldn't even begin to enumerate the range of services she has provided, not just in helping produce the book but in supporting every conceivable aspect of my professional work. Like Jan, Virginia is astonishingly patient and kind. I'm not sure how I came to the good fortune of having these two people in my life. Sometimes the dice roll in our favor.

Last but far from least, many thanks to my students at the University of Connecticut School of Social Work, whose enthusiasm for the first edition of this book has made it all worthwhile. This book has truly been a community multicultural multimedia project. Thanks and apologies to anyone I may not have mentioned here. All errors, of course, are my own.

Harriette C Johnson, June 2004

CONTENTS

Forewords to the First Edition

The central feature of contemporary research on mental illness is a focus on the brain. It follows that a most pressing need of both brain research and its application to clinical problems is effective communication across the many levels of analysis and the many disciplines involved. Effective communication is all too infrequently the practice, however. Many scientists tend to work, present, and publish within narrowly defined fields, while many people whose interests and expertise lie outside of science assume that until someone takes care of translating new knowledge into usable information, the processes and objectives of research are, if not irrelevant, at least pre-relevant.

Fortunately for all of us—scientists, service providers, patients, and anyone whose tax dollars or personal contributions support research—Dr. Harriette Johnson does not make that assumption. As a professor in the School of Social Work at the University of Connecticut, she can claim the rare perspective of one who stands atop the divide that so often separates research from practice. She is an avid student of neurobiology and related sciences and, in what is by far the largest of the mental health clinical professions, an engaging and talented teacher of that same subject matter.

I first met Dr. Johnson while I was conducting a workshop on the molecular foundations of addiction at Harvard University in 1994. At the conclusion of the session, she flattered me by requesting copies of several slides I had used. I agreed readily and thought no more of it. Shortly thereafter, however, I opened a piece of mail to find copies of the teaching slides I had lent out, but wonderfully emended to meet the needs and interests of audiences who, though not as fully immersed as I, nonetheless are intrigued by what research has to say about the roles and processes of neurotransmitters, receptor subtypes, DNA transcription, neurocircuitry, and other simple facts about drug addiction. I was impressed and must confess that I used those modified slides on more than one occasion.

Since coming to the National Institute of Mental Health, I have been made increasingly aware that breaking down barriers of communication has never been more important. We are moving into an era in which we are seeing the long-sought merger of molecular approaches and integrative neurobiology, conceptually as well as in our actual lab and clinical work. In this era, our scientific success and, ultimately, the well-being of people with severe mental illnesses and other brain disorders, hinge on our ability to speak to and work with people in disparate fields of specialization. As I have no doubt that achieving the necessary communication and collaboration will prove to be a formidable task, I am pleased to find that we at NIMH are not alone in our attempts to address the issue.

What Dr. Johnson demonstrated first with a few of my slides, she now has repeated in this eminently readable and utterly fascinating monograph, *Psyche, Synapse, and Substance*. Dr. Johnson clearly recognizes that the ongoing revolution in molecular science, and, paralleling it, our increasing capability to investigate brain function at an integrative level, are now making it possible to weave together the influence of genetic, developmental, and environmental factors. This integration is yielding a picture of human behavior, including mental illness, of unprecedented clarity.

The goal of NIMH-sponsored research is more effective diagnosis, treatment, and, eventually, prevention of mental disorders. With each advance, the science of our field is having an increasingly immediate and greater impact on how we diagnose and treat mental disorders. Thus it is imperative that all professionals in the field, all advocates, and all citizens appreciate and benefit from the knowledge and insights of others. *Psyche, Synapse, and Substance* lucidly and comprehensively makes available to a diverse readership one of the most powerful and promising knowledge bases to come out of 20th century medicine. A broad appreciation of that knowledge base will do much to ensure that the power and the promise will be realized even more fully in the 21st century.

Steven E Hyman, MD, January 1999
Director, National Institute of Mental Health, 1996-2001
Provost, Harvard University, 2001-present

✱✱✱✱✱

I can think of no area of knowledge integral to social work practice that has experienced greater growth than neurobiology. One of social work's major strengths is its focus on the person-in-environment, which recognizes the interplay between human biology and the social environment in shaping emotions, behavior, and cognition. By and large, however, the social work curriculum has had difficulty fully incorporating up-to-date knowledge about brain functioning. Partly this is due to the tremendous knowledge explosion in this "decade of the brain"; partly it is due to the complexity and highly technical nature of the subject; and partly it is due, I fear, to resistance based on the false assumption that a neurobiological understanding of psychic phenomena leads to a biologically deterministic view of human emotion, thought, and behavior, and is, therefore, antithetical to psychosocial interventions.

These impediments to incorporation of the new knowledge base into social work curricula are what make *Psyche, Synapse, and Substance* so valuable, because it directly addresses these areas of concern. It presents cutting-edge neurobiological knowledge, based on recent research, in a highly readable, informal style, without condescension and without compromising the book's scientific integrity. It also relates this basic scientific knowledge about how the brain works to human psychological development and psychosocial interventions. Consequently, this wonderful monograph will serve especially well as a supplemental text for both graduate and undergraduate courses in human behavior and the social environment, mental health, substance abuse, and related areas. It should be required reading for all practitioners concerned about health, mental health, and human development.

This is an especially exciting time for professionals concerned about understanding and treating psychosocial problems as well as for those developing social programs and formulating public policies dealing with a wide range of issues, including health, mental health, child development, and criminal justice. Almost daily, new discoveries are announced by molecular biologists, geneticists, and other scientists studying human biology. Many of these findings have direct implications for both public policy and psychosocial practice. To remain relevant and effective, social work educators and practitioners must stay abreast of this knowledge explosion. It is, of course, not possible to do so broadly without the assistance of scholars who will synthesize and translate this information into usable form while remaining faithful to the scientific integrity of the new research-based knowledge.

Psyche, Synapse, and Substance is a splendid example of this synthesis and translation. We need more of it.

Donald W Beless, MSW, PhD, January 1999
Executive Director
Council on Social Work Education, 1988-2003

Preface to the Second Edition

The intention of the first edition of this book was to present fundamentals of brain/behavior relationships that social workers need to know. I attempted to convey an integrated biopsychosocial understanding of human psychological events, and to demystify roles of the brain in emotion, cognition, and behavior. These basics of neurobiology are deemed essential by most of the scientific community for understanding human psychological functioning, child and adult development, mental illness, substance abuse, and addiction (Health and Human Services, 1999). The practice community too is increasingly recognizing the importance of this knowledge for our work in the front lines.

The audiences for this book are students in nonmedically oriented helping professions (using the book as a course text), and practitioners in several helping professions (using the book as a self-teaching tool). As a social work educator, I often refer to applications in social work. However, the book's subject matter is equally relevant for non-social work practitioners such as guidance counselors, family therapists, psychologists, and rehabilitation counselors.

The second edition differs significantly from the first in that the scope of the book has been greatly expanded. The second edition is about twice the length of its predecessor, because the focus of the first edition, fundamentals of neurobiology, is now only one of several equally important sections of the book. What were very brief chapters on child development, mental health/illness, and substance abuse/addiction in the first edition are now more fully developed. The new sections report cutting-edge research pertinent to each area of inquiry. Practice-related vignettes have been added in several sections. I have also added a module on multiple routes to quality of life such as spirituality, exercise, and herbal remedies.

Words that characterize the book's perspective are very familiar to social workers: biopsychosocial, ecological, multisystemic, and, of course, focus on person-in-environment. I summarize knowledge pertaining to biological and nonbiological risk and protective factors, best practices, evidence-based approaches, and interventions that show promise but have not yet been fully validated by research. It's important to note that *biological* risk and protective factors can emanate either from the individual (e.g. as a configuration of specific genes) or from the external environment (e.g. as viruses or toxins). When I refer globally to environmental influences, I mean forces from outside the individual that include biology as well as nonbiological influences.

I have updated and expanded explanations of fundamental mechanisms by which major classes of drugs are believed to work in the brain. Equal attention is given to drugs designated "therapeutic" (medically sanctioned) and drugs for pleasure, referred to as "recreational" (sold on the street or, in the case of alcohol, tobacco, and coffee, in liquor stores, supermarkets, and convenience stores). Our purpose is to educate practitioners about some of the ways these classes of drugs work, so that they in turn can educate individuals and their families, work with physicians and other medical personnel as credible partners rather than as handmaidens, better advocate to meet therapeutic needs of individual clients, and, in the role of program administrators, improve the quality of the agency practices they authorize and shape. The book is not intended as a compendium of therapeutic and side effects of specific drugs; manuals that compile this information are available elsewhere.

The major topics of the book have been updated and expanded in the following ways.

Biopsychosocial knowledge (Part I). I have added the epigenetic model of biology/environment interaction, in the tradition of complex adaptive systems. The discussion of theoretical frameworks and unifying themes has been developed and expanded. Because the environment is so critical for most psychological events, a practice example attempts to show how to begin to identify complex, recursive interactions between the brain, the individual, and the environment.

I emphasize the word "begin." As Steven E Hyman, recent director of the National Institute of Mental Health has said (1997), "the brain is the most complex entity in the known universe." It would be the height of folly to suggest that assessments involving such complex factors, many of which are not yet well understood, could be conclusive or certain.

Fundamentals of neuroscience (Part II). Information has been updated and considerably expanded.

Child and adult development (Part III). The topics continued from the first edition (temperament, stress, resilience, critical periods in development, affiliation, and attachment), have been updated and elaborated in more depth. I applaud what science has been able to uncover, and at the same time caution readers that much of our presumed knowledge is provisional, awaiting validation, and subject to revision or outright refutation by future research. I have added a new chapter on consciousness.

Major psychiatric disorders (Part IV). In the first edition, neurobiological aspects of several conditions were summarized briefly. In the second edition, these brief reviews have been replaced with a discussion of assessment issues and a biopsychosocial model of assessment that incorporates neurobiology and psychopharmacology, cultural and organizational influences, and client preferences and priorities. Issues such as concepts of normality, diagnosis, and influence of professional cultures are considered. Case material illustrates professional dilemmas and potential conflicts between organizational imperatives and client needs.

We then look at implementation of the complex adaptive systems perspective described in Part I. To demonstrate use of the model, connections between emotions, behavior, and cognition and their neurobiological correlates are presented with respect to one DSM diagnosis, borderline personality disorder (BPD). The assessment process for BPD can be applied across the board for emotional, behavioral, and cognitive challenges that have been subsumed in diagnostic categories. A detailed social work application of the model follows.

Substance abuse and addiction (Part V). Material from the first edition has been updated and expanded, and research pertaining to effectiveness of interventions has been added. This section presents some known explanations for the workings of drugs in our brains and our psyches, how practitioners can use this knowledge to help clients, and some provisional beliefs about addiction that are still in the process of testing.

Multiple Routes to Quality of Life (Part VI). This section is new in the second edition. We substitute the words "multiple routes to quality of life" for the commonly used expression "alternative therapies" because we include approaches that are usually not designated as therapy. The therapeutic value of touching was discussed briefly in the first edition. Now we also consider spirituality, work, exercise, pet therapy, and herbal medicine. I have the utmost respect for people's ability to improve the quality of their lives by approaches that sometimes have not been as fully appreciated by professionals nor as diligently studied by scholars as they should be.

The book combines short chapters of text with accompanying fill-in and short-answer questions as well as anatomical drawings to be labeled. For the dual purposes of class instruction and self-teaching, all books are sold with question sets, which are now in a separate workbook rather than interspersed throughout the text itself. These exercises reinforce the reading in small increments, to avoid overwhelming readers. First edition readers have reported that the reading and exercises are fun to do, and helpful in important ways for their practice. Because new materials are available (animated graphics, practice-related vignettes, and expanded content on human development, psychiatric disorders, and addiction), I believe that the second edition will engage and enrich readers even more.

Since few social workers have studied neurobiology, I have attempted to explain some essentials of neuroscience in everyday language. The book's style is conversational, sometimes even playful. Although some scholars may prefer a more formal style, feedback from many of the social work programs that have adopted the book, as well as from our own students, has convinced me that conversational style, simulating a person-to-person spoken dialogue, helps to bridge the gap between our people-oriented readership and the scientific, research-based content of the book.

A CD-ROM has been produced that illustrates the text with multimedia animated graphics. These animations enhance the neuroscience graphics in the book. The CD-ROM is easy to use on most personal computer (PC) and Apple computer (Mac) platforms, and is available to instructors in conjunction with adoption of the text.

Detailed explanations of practice methods—other than those related to biological aspects of practice—are beyond the scope of the book. I do identify specific worker behaviors for integrating biological knowledge into practice: accessing state-of-the-art information pertaining to client concerns; educating clients about research-based knowledge; helping them access this information, understand it, and relate it to their own issues; facilitating and supporting clients' independent decision-making skills; locating resources for services; disseminating information to colleagues; advocating for agency policies consistent with best practices; and maximizing political effectiveness to bring public policies more in line with the best interests of service users.

To the best of my knowledge, this book is the only social work text to compile neuroscience information for practice in a systematic form. To present essential neuroscience knowledge, and then to try to integrate it with the psychological and social knowledge that our profession has used in practice for many decades, is a daunting task, and I am only making a beginning here. The current state of knowledge across disciplines does not allow any of us—social workers, psychiatrists, or neuroscientists—to make claims beyond a beginning understanding of these processes. Neuroscientists are further along the road toward understanding how the brain works; social workers (we would hope) are further along the road toward understanding people's real life contexts. Our various domains of expertise must be combined. Each discipline needs the expertise of other disciplines. I hope that practitioners, students, and scholars will find these fundamentals useful for applying, building on, and advancing our knowledge beyond this beginning work.

PART I

BIOPSYCHOSOCIAL KNOWLEDGE FOR SOCIAL WORK PRACTICE IN THE 21ST CENTURY

1
The Biopsychosocial Perspective: Introduction

In social science, the nature/nurture controversy has shaped causal models of behavior. . . . .now, [we] will need to be trained in the new person-environment paradigm, which includes the integration of nature and nurture. . . .[We] need to be open to new concepts. Do not fear integration of social and physical science, but welcome it. This is the future.

AM Johnson and S Taylor-Brown
Social Work Education Reporter, 1997

Are biology and psychology inseparable? For many decades, this notion was not seriously entertained by purveyors of psychological wisdom. It was assumed (without proof) that psychological disorders were either "organic" (as in senile dementia or drug-induced psychosis) or else "functional," a term that denoted environmentally induced states of mind. Nearly all mental disorders and emotional difficulties were assigned to the functional category. The word "environment" itself was shorthand for psychosocial environment, roughly translated as "the people in my life that affect my psyche." More specifically, with respect to emotional and cognitive development, "the environment" was assumed to consist of parental behaviors, feelings, and thoughts continually bombarding developing offspring.

Few mental health professionals in the 1950s and '60s seemed to have difficulty with such puzzling questions as how a disembodied mind might actually work. It didn't seem important. What social workers, psychologists, and psychiatrists did

3

deem important were parental sins of commission and omission (mostly by mothers) that led inexorably to emotional damage in their young (Caplan and Hall-McCorquodale, 1985). Clinician/detectives were charged with finding these pathogenic forces responsible for their clients' unhappiness.

Today, few scholars in the fields of child development, mental health, and related areas still accept the mind/body dichotomy, i.e. the separation of mind from body. The so-called "biopsychosocial paradigm" is now a cornerstone of behavioral theory. Yet many in the helping professions have been slow to learn the scientific knowledge base that is required to meaningfully integrate "bio" knowledge with the more familiar "psycho" and "social" components (Beless, 1999). As a result, educational curricula in some of the nonmedical helping professions have not incorporated the "bio" into the biopsychosocial model endorsed by arbiters of professional education (Council of Social Work Education, 1995; Biggar and Johnson, 2004).

It should not be surprising that professional knowledge has not kept pace with neuroscience's recent illuminations of brain/behavior relationships. These advances, made possible by technological developments in brain imaging, biochemical analysis, electrophysiological measurement, and other procedures, have been so monumental and so swift that they have been described as "explosions," "avalanches," and "revolutions."

Another obstacle to embracing the new integrated person-in-environment model is that biology has been a vehicle for oppression, linked to characteristics such as skin color, facial characteristics, gender, and ethnicity. For example, the horror of Holocaust genocide, enslavement of African Americans, racially motivated hate crimes, imprisonment of Japanese Americans in concentration camps, victimization of Native Americans, and blaming women victims of physical and sexual abuse for their abuse, has led many social workers to regard biology as the source of the evil, and therefore to deny its role in psychological phenomena. Yet clearly these atrocities are not caused by biology itself; they are social psychological responses to apparent or actual biological differences (Aronson, 2004).

The first edition of the book emphasized universal neurobiological phenomena—events similar for all human beings regardless of race or ethnicity. This content is retained in the second edition. We consider information that helps us begin to understand *how societal goods or evils actually affect our brains and minds* to give us strength, pleasure, fun, or serenity, or to cause us stress, anxiety, conflict, or trauma. How are influences in childhood or adulthood experienced by our brains and translated into psychological events? How do individual differences interact with environmental forces to mold our lives? How can we make connections between the conditions we encounter every day in our work, such as substance abuse, psychiatric challenges, or family violence, and the invisible changes in our brains that lie behind these visible psychological events? The more we can learn about these connections, the better equipped we'll be to help people we serve.

Emotions, behavior, and cognition are processes that take place at the nexus where the individual human organism ("person") interfaces with the world outside that individual ("environment"). It's important to remember that the biology/environment dichotomy itself is not accurate, because environments can be biological as well as psychosocial. For example, viruses and bacteria (which sometimes cause mental illnesses), or meats and vegetables laced with environmental toxins, are biological aspects of our environments. (Most of the food we eat is organic material, with or without added inorganic compounds such as preservatives or colors.) Physical nonbiological aspects of the environment such as heat, cold, dryness, light, and darkness also influence our moods and behaviors. The point is that we need to expand our concept of environment beyond the "psychosocial" to encompass biological and nonbiological physical conditions or agents.

Brain/psyche connections are important in enhancing our understanding of typical (so-called "normal") human development, interpersonal relations, and human behavior in the social environment. In this book I try to avoid the term "normal" because of its cultural relativity. "Typical" and "atypical" are used instead, to convey the notion of frequent or infrequent occurrences represented on a continuum (bell-shaped curve), without connotation of judgments such as "abnormal," "pathological," or "dysfunctional"—judgments that tend to reflect perceptions of the beholder.

Since the publication of the first edition (1999), research on psychiatric disorders, substance abuse and addiction, and child and adolescent development has continued at an extraordinary rate. It is now known that there is no single cause for most psychiatric disorders and most instances of addiction. With rare exceptions (such as Huntington's Chorea), these conditions and states arise from complex multifactorial interactions through time between *risk factors* and *protective factors*. Here, risk factor is defined as a biological or nonbiological variable within or outside an individual that contributes to a problem occurring, makes the problem worse, and/or makes the problem last longer. The converse, protective factor, is defined as a biological or nonbiological variable within or outside an individual that contributes to preventing a problem from occurring, lessens its severity, and/or helps it get better faster.

Major risk factors and protective factors, both for illnesses and for social problems, have been identified through epidemiological, clinical, and laboratory research. Risk and protective factors arise from genetics, viruses, birth events, nutrition, level of economic advantage, level of social supports, and the salience of cultural forces. Examples of cultural forces are ethnocentrism (for example, perceiving one's own ethnicity as superior, or imposing beliefs derived from one's own cultural background on persons of other ethnicities); heterosexism (perceiving people with same sex preferences as inferior, pathological, or not entitled to equal protection under the law); or ableism (perceiving persons with disabilities as alien, incapable, or equivalent to the disability itself rather than an ordinary person who has a disability). Many other factors that can impinge on individuals and families through time have been identified in research literature as risk and protective factors.

Imagine two piles on a balance scale. These forces may be in the risk factor pile, the protective factor pile, or both at various times. For example, closeness of family ties may create strong social supports for a person with a mental disorder, but also generate troublesome conflicts between that person and a loved and loving relative. As another example, having many friends may provide a substance abuser with support and a sense of belonging, but these friends, if users themselves, may at the same time reinforce use and deter abstinence. Moreover, appraisal of risk factors and protective factors requires us to go beyond two "piles" and evaluate the transactions across the piles, not just in the present nor at some moment in the past, but continuously through time.

Thus the metaphor of two piles, one of risk factors and the other of protective factors, must be replaced with a more dynamic, interactive image. A boxing match comes to mind. Imagine that one combatant represents "Risk Factors," the other "Protective Factors." You watch a continual series of split-second actions—footwork, body motions, and jabs back and forth—as each adversary responds to the other's action. This rapid non-stop action continues until one warrior knocks down the other. If the Risk Factors prevail over the Protective Factors, an observable condition or crisis occurs. If the Protective Factors prevail by fending off the onslaught of the Risk Factors, no identified disorder emerges and/or an impending crisis is averted.

During this entire process, let's imagine that you are a participant-observer, and that this drama is taking place inside your own brain and body. The risk and protective factors in the contest are your own. Your mental and physical state is the winner's prize. Your brain processes every step the combatants take, every facial expression, every punch. Your brain is doing high-speed calculations about the effects of one protag-

RISK FACTORS   PROTECTIVE FACTORS

onist's latest move, the likely next move of the other, the effect every move has on your own physical and mental state, the sensations you experience with every move of the combatants, and how you should respond to improve your odds of winning.

Yet the boxing match image also falls short of representing reality by combining all the risk factors and the protective factors into single entities, the two boxers. To try to make our metaphor more realistic, we make each risk and protective factor a separate combatant. The boxing ring then morphs into a giant arena with many players, each interacting with several of the others, in their own camp (risk or protection) as well as in

the opposing camp. Some players act simultaneously as agents of both risk and protection. In the example above, we noted that when an individual is an addict, user friends supply social support and friendship, but at the same time (inadvertently or openly) undermine the individual's attempt to get clean.

The risk factor/protective factor model, developed in the field of epidemiology, pertains specifically to assessment of physical illness vs. health, and social illness (such as crime, violence, danger) vs. social well-being (such as order, civility, safety). That is, it implies a concept of physical or social pathology. Of course, in the entire field of human behavior in the social environment, pathology/wellness is only one of several dimensions. That is, there are various ways of viewing social phenomena beyond making judgments about their inherent sickness vs. wellness, or goodness vs. badness.

Social work's human behavior knowledge for practice is systemic, broad in scope, and unimaginably complex. This knowledge pertains to space and time; hierarchies of size (micro, mezzo, macro); level of organization; mobility; capacity for change; accessibility to human comprehension; cultural and ethnic difference; degree of power; access to resources; propensity for wellness and for specific illnesses; temperamental attributes; multiple neurobiological and muscular endowments in areas of human endeavor such as music, athletics, visual arts, mathematics, language, information processing, mechanics, and many other abilities.

Let's try another metaphor here. Imagine that the many variables influencing these components are flakes in a dried potpourri of herbs and seasonings. When we toss the potpourri into a cauldron of simmering stew, the flakes jiggle and rock around the pot, become limp as they absorb the liquid, change their own form and perhaps their chemical composition, and at the same time saturate the vegetables that are cooking in the pot. By the time cooking is complete, the flavor of the food in the pot has been changed, to make it more peppery, sweeter, saltier, more bitter, or altered in an infinite number of possible ways. Our basic carrots, greens, and potatoes are still there, but they taste so much more lively, delicious, or, on the other hand, perhaps more unpleasant, than they would have tasted in their simple boiled state. We'd like to understand something about the recipes that produced the flavors that tasted good or bad to us so we can improve the stew next time we make it. We really have no idea about the molecular processes that led us from the original foods to the outcomes we liked or didn't like (nor do we need or want that information). What we do want is some basic knowledge about how the seasonings we add produce the delicious versus the distasteful stews. Which elements of the potpourri enhance which vegetables, meats, or fish, at what temperature, amount of added water, duration of cooking, and sequence of adding ingredients?

In our stew, there are quite a few known basic foods and quite a few known seasonings that we start off with. When we act as culinary consultants in the kitchens of our clients, there are many more variables, many of them unknown, and many more complex dynamics and varying processes than in the creation of our stew. In our assessments, how do we make a coherent whole out of the many variables interacting with each other in this biopsychosocial stewpot? How do we avoid becoming overwhelmed by this complexity?

Just *recognizing* the complexity of human psychological functions in their environmental contexts is the first step. Perhaps the second step is to acknowledge that in any situation there will always be factors operating that we're not aware of, and that our understanding will always be imperfect. The myth that we can "master" a content area such as human behavior in the social environment or mental health must be dispelled.

This book comes in on the third step. As we increasingly recognize the complexity of these areas of knowledge and then acknowledge how little any of us knows, we can begin to identify specific kinds of knowledge that can help us in our work of helping others. We can recognize that we need to know more, especially in certain areas, than we had previously believed. The neurobiological foundation of human psychological functioning is one kind of knowledge we once ignored. Now we're stepping up to the plate and informing ourselves.

How can this book help? The purpose of the book is to make available some knowledge about the role of neurobiology in psychological functioning that can be integrated with psychosocial and life contextual knowledge. How often in our everyday lives do we unconsciously integrate knowledge from different areas into our assessments? Wouldn't it be preferable to be more aware of knowledge that we're integrating? What do we do when an accurate assessment requires some expertise or piece of information that we don't have? Do we pause in the assessment to try to obtain that information? What happens if we *don't even know* that we're missing an important piece of information? Is our assessment off target?

Yes, probably. Owing to the complexity of human situations, most of the assessments we make continually every day, knowingly or unknowingly, in our work and our personal lives, are probably less than fully accurate. But in situations with clients, we want to reduce inaccuracies as much as possible while recognizing the limits of our access to important information.

As you proceed through the book, you'll learn about some neural pathways that transmit critical information, some brain systems that mediate psychic processes, and functions that brain structures perform as they transmit messages using chemical messengers called neurotransmitters (the process of neurotransmission). You'll gain information about how forces in the environment can interact with the brain, and you'll become familiar with recent discoveries about critical periods in child development.

You'll discover how the brains of people diagnosed with various psychiatric disorders may differ from each other and from typical brains. You'll learn some ways prescription and street drugs act in the brain, and you'll become familiar with some of the changes that can take place in the brain as people become addicted to a substance.

Does a neurobiological understanding of psychic phenomena explain away the roles of emotions, behavior, and thought? *Not at all!* Does it take away the need for psychosocial interventions? *Not at all!* On the contrary, it helps us understand psychic phenomena (mental functioning) more fully so that we can integrate this knowledge with our psychosocial expertise.

This second edition (2004) attempts to incorporate highlights of the information derived from a knowledge base that has grown exponentially during the past five years, in both scope and depth. Thousands of publications in scientific journals can be accessed through databases such as PubMed. The sheer volume of available information, not to mention the technical complexity of much of the research, might tempt us to throw up our hands. Do we really need to be conversant with all this? Surely there's no way most of us can understand much of the research on genetics or the neurobiology of medication actions, and do we really have to?

Here, my response is both yes and no. No, we don't need to understand advanced scientific treatises on genes and alleles, and no, we don't need to master much of this highly technical information to help our clients. So where's the "yes"?

The "yes" is that our service users do need to understand what their own or their family member's condition is, how it can affect behavior and emotions, how the condition may manifest itself in the future, and what the new knowledge about these issues may mean for themselves and their loved ones. They need to learn how a specific treatment or intervention is thought to work, and what interventions have been shown to be effective for some people with similar conditions. That is, clients need to have at least a basic understanding of the "hows" and "whys." How is treatment X or Y likely to address the problems of concern, and why? How can they make informed choices? And on a more existential level, clients want answers to questions like "Is my situation hopeless?" "What did I do to cause this problem?" "Am I just a loser?" "Am I a bad parent?" "Did I fail as a partner?" "What can I do to make things better?" "Who can help me?"

Is it the social worker's responsibility to facilitate clients in making informed choices and resolving existential issues? If so, then we as professionals need to understand *enough* about the way the condition operates (biologically as well as psychosocially) and *enough* of how possible interventions work (biologically as well as psychosocially) to convey *enough* of that information to clients so they can better understand, learn, and take action.

This book is intended to help you acquire some tools, whether you have a limited scientific background or are well versed in the biological sciences. The book's style is conversational, but its message is always serious. The research from which the author has drawn is referenced at the end of sections, but sometimes, in order to enhance readability, it is not cited in the text itself. The book provides information essential for course content related to human behavior in the social environment, mental health/mental illness, substance abuse/addiction, child development, actions of therapeutic and street drugs in the brain, and practice interventions relevant to these areas.

2
What Does All This Information Have To Do With Social Work?

If you're a practitioner, you may be wondering why you need to know this content in order to help clients who are suffering from effects of poverty, disease, family problems, addiction, or violence. Probably interesting, you say to yourself, but a luxury I have no time for. I have to be out there every day struggling to help real people with real problems. Those academics aren't in the front lines. They can afford to learn interesting but irrelevant information. I can't.

But can you afford not to? No! This knowledge does directly affect the well-being of those real people you're trying to help, and it can directly affect how you practice. Parts III, IV, V, and VI summarize some current neurobiological research on child development, psychiatric conditions, substance abuse/addiction, and multiple routes to quality of life. Those chapters will familiarize you with evidence that has led neuroscientists and medical researchers to their conclusions. In this chapter, I'll try to make the case that social work practice needs this knowledge base.

A brief history of beliefs about mental disorders can help us understand how we've come to require more knowledge. In the early decades of the 1900s, most psychiatrists in Europe and America believed that mental disorders were illnesses of the brain. By the 1950s and '60s, this view had been largely supplanted by new ideas proposed by Freud and his followers. Mid-20th century psychiatric dogma, along with social work's conventional wisdom, now categorized all mental disorders either as "organic" (biological in origin, such as senile dementia, drug intoxication, and mental retardation) or as "functional" (arising from psychosocial stressors). I'll refer to this formulation as the "dichotomous theory of etiology" (see Professional Beliefs about Causes of Mental Disorders: A Quick History).

Professional Beliefs about Causes of Mental Disorders: A Quick History

BF (Before Freud): Physicians believe that schizophrenia and other mental disorders are biological diseases of the body organ called the Brain.

Early 1900s: Pavlovian stimulus-response theory gets around.

 1920s: Freud comes to America. His views sweep the country.

1920s to 1960s: Dichotomous Theory of Etiology captures psychiatrists and social workers.

Organic
Senile Dementia
Drug Intoxication
Mental Retardation

Functional
Everything Else

The **Everything Else** means that the adversaries in the **War of the Inner Drives against the Voice of Reason**, carried out on the **Battlefield of the Psyche**, do not reach a **Peace Agreement**.

The result? **Mental Disorders!**

Mid-20th century: Skinner launches operant conditioning. Are mental disorders *learned*?

1960s, '70s, and '80s: Family Systems theorists promote a different concept: Mental disorders are manifestations of Dysfunctional Family Systems. They are not real mental illnesses inside one individual. The "ill" family member turns out to be a "symptom carrier" for the entire family.

Dysfunctional Family System

Which Family Member is the Symptom Carrier?

1970s, '80s, and '90s: Behaviorists broaden their thinking to embrace Cognition and Emotion (the birth of Cognitive-Behavioral Psychology), and Neurobiology.

1980s and '90s: Advances in technology spawn new experts on mental disorders, the Neuroscientists. What has their research shown? Those BF (Before Freud) physicians were right after all. Mental disorders are physical illnesses of one of the organs in the human body, the human brain!

In the decades before the 1970s, every psychiatric disorder not in the organic category was automatically assigned to the functional category, from autism to schizophrenia to manic-depressive psychosis (as bipolar disorder was then called) to character pathology (now known as personality disorder). Cognitive differences, too, were often viewed as functional. For example, what we now might call a "learning disability" was then thought to be a learning problem caused by emotional factors. Every condition not designated organic was assumed to be functional, i.e., caused by toxic environments, most typically rejecting, overcontrolling, overindulgent, distant, or otherwise inadequate mothers. Fathers were seldom mentioned in those days.

In the 1970s, tiny cracks appeared in this fortress of beliefs, as a few people questioned the popular concepts of the "schizophrenogenic mother" (assumed to have caused her child's schizophrenia) and the "refrigerator mother" (Bruno Bettelheim's characterization of mothers of autistic children). Bettelheim (1903-1990) also likened mothers of autistic children to SS guards in Nazi concentration camps. He continued proclaiming his views to sellout audiences of social workers and other service providers until his death.

In the late 1950s and early '60s, schizophrenia began to be treated with Thorazine and other antipsychotic drugs. A large number of long-term sufferers improved so much with these medications that psychiatric hospitals discharged them after years of hospitalization. In the 1970s, early neuroscientists proposed that schizophrenia, autism, and bipolar disorder were neurobiological illnesses in the organ of the brain. Studies by Kety (1979) and others supported the importance of genetics in schizophrenia, noting that most identical twins of persons with schizophrenia either also developed schizophrenia (about 50%) or had other mental disorders along a "schizophrenia spectrum." Fieve (1975) found rates of genetic influence in manic-depressive disorder to be even higher than those for schizophrenia. Cantwell (1972), Wender (1971), and others documented strong genetic and other biological influences in minimal brain dysfunction, a global term that encompassed what we now call attention-deficit hyperactivity disorder (ADHD) and learning disability. Ritvo and colleagues (1976) demonstrated multiple brain abnormalities in children with autism. Lewis and Balla (1976) went so far as to obtain neuro-logic data suggesting that brain dysfunction often played an important role in delinquent behavior. By 1981, major biologi-cal substrates of borderline personality disorder had been iden-tified by Andrulonis, Glueck, Stroebel, and colleagues.

These prophets of science were largely ignored by mental health practitioners and educators. At most, they conceded that there might be a "biological predisposition" to these disorders. Still, many held tenaciously to the notion that toxic family envi-ronments (read "bad mothers") made the decisive contribution both to major mental illnesses such as schizophrenia and bipo-lar disorder, and to nonpsychotic conditions such as obsessive-compulsive disorder, panic attack, and mood dysregulation—conditions that often inflicted misery on sufferers and their fam-ilies (see Part IV, Major Psychiatric Disorders).

The convergence of new methods for studying mental ill-ness and the formation of the National Alliance for the Mentally Ill (NAMI) led in the 1980s to widespread questioning of such concepts as the schizophrenogenic mother, the refrigerator mother, and the idea of mental illness as a myth of attribution thought up by culpable families to cover up their dysfunctional family systems. But research evidence increasingly supported the view that the origins of learning disabilities and ADHD were neurobiological, not responses to parental ambition, mar-ital conflict, or other family dynamics.

Brain-scanning techniques, developed in the 1980s and per-fected during the 1990s, brought the history of theories about mental disorders around full circle to the beliefs of the pre-Freudians. Leaders in the neuroscience revolution now asserted unequivocally that psychiatric disorders were biological dis-eases of the brain, just as nephritis is a disease of the kidney. They claimed that these neurobiological diseases were not caused by toxic families any more than nephritis is caused by toxic families. Unlike the pre-Freudians, however, these neuroscientists had masses of scientific data to support their claims.

But there was a problem. Thousands of practitioners had been born, raised, and immersed in the old dichotomous mind vs. body paradigm. They had spent years of their lives developing treatment strategies and techniques founded upon that model. Now they were being asked to give up a lot of what they had been taught and to learn to think about things differently.

To make matters worse, the domain of neurobiological research extended further and further beyond its original focus on major mental illnesses to encompass what had once been perceived as the turf of nonmedical psychotherapists—nonpsychotic conditions mentioned previously, addiction, and personality disorders. One by one, each diagnosis was removed from the "functional" column and placed in another category—not the old "organic" column, but an entirely new category dubbed "neurobiological disorder." This new designation assumed that persons suffering from any of these conditions had an underlying, biologically based condition that continually interacted with environments through time, with the biological organism influencing the environment and the environment influencing the biological organism, back and forth, in endless flux. In fact, the time had come to deep-six the old organic/functional classification system.

The new knowledge generated by the neuroscientists called for major shifts in practice approaches for people with mental illnesses, including those with less severe conditions that could be subsumed under the rubric "psychological problems," or those with deficits or differences in cognitive processing, such as learning disabilities. Rather than simply helping people gain insight into their emotions and ventilate their feelings, therapists were now called upon to use a range of interventions for clients with diverse diagnoses, usually including a combination of some of the following: education about the condition; medication; training in social, communication, self-control, or daily living skills; support groups; special educational strategies;

individual counseling; and active ongoing advocacy to help clients obtain employment, find housing, or access other financial, material, and service benefits. The one-time mainstay of psychiatric treatment, psychotherapy, was now supplemented with many other interventions. Educational supports and techniques targeting specific neuropsychological functions replaced psychological counseling as first-line interventions for learning deficits.

In addition, psychotherapy itself now had to take a variety of forms, depending on the condition. A client with schizophrenia might require education, supportive counseling, and coaching about how to be successful in tasks of daily living. Cognitive restructuring might be indicated for depression, in vivo exposure therapy for acute anxiety, self-control skills training for impulsivity problems, or relapse-prevention techniques for addiction.

The simpler days of sitting in an office helping clients express their feelings were over. A much wider repertoire of techniques was now required. This expanded menu of interventions was consistent with the findings of the new neuroscience, and these therapeutic techniques could be used in conjunction with medication or without it.

By the late 1990s, scholars had achieved a deepening understanding of all psychological conditions as outcomes of the interactions between biological and nonbiological risk and protective factors. Psychological distress or dysfunction was not, as previously assumed, an inevitable consequence of a single variable, BP (Bad Parenting). In fact, persons with psychological or cognitive problems, whether severe or relatively mild, often had families who were concerned, loving, and persistent in their efforts to help their loved one. These parents, who usually had other perfectly typical children, were remarkable not for their pathogenic qualities but for their dedication to the needs of their ill or challenged family member.

Parents who were angry, critical, depressed, or in denial about their child's condition were now seen not as pathogenic agents, but as deeply worried and sad people who felt helpless to make their family member better. They were experiencing vicarious pain for the child's seemingly endless suffering, chronic grief about the loss of a child they once had or hoped to have, or they felt overwhelmed by crushing life circumstances.

Professional helpers now increasingly subscribed to the view that all parents, including the most supportive, can certainly do things that make their loved one's condition worse— not as pathogenic agents but as people in need of emotional support, validation, and specialized skills for relating effectively to the challenged family member. Accordingly, as the persons with the mental illness or psychological distress were now known to need a range of interventions, family members also were found to need professional services quite different from what social workers and other service providers were accustomed to offering.

Okay, you say, but I still don't see why I need to learn about the neurobiological basis for psychic functioning. Perhaps some practice questions will help you understand the "why."

Suppose a teacher refers you a five-year-old boy because he's hyperactive in class. How can you determine if he's just one of those many kids who are often restless, impulsive, and overactive? Or is his hyperactivity a reaction to a stressful family or school situation? Or does he have a neurobiological condition called attention-deficit hyperactivity disorder? How will the answer to that question affect your ideas about intervention? Will a behavior management plan do the job? Should the family situation be explored to see whether there are some acutely stressful events taking place? Is medication likely to be needed?

Here's another example. What will happen when your client, or a family member of your client, asks you to explain what's wrong with her? Or asks you how a particular medication works? Your client says that unless you can explain to her what this drug is supposed to do and how it works, she's going to stop taking it. Why should she risk putting this powerful (and mysterious) chemical into her body? Will you shrug and say she'd better talk to her doctor? You work with this doctor, and you know he doesn't take the time to give more than a cursory explanation. He's too busy.

What will happen when you have a strong opinion that your client's medication isn't doing what it's supposed to? When the psychiatrist asks you why you think so, and you don't understand how the medication is expected to work in the first place, what are you going to say? How credible will you be in the eyes of the prescribing physician who has so much power over your client's well-being? What basis will you have for making judgments about the course of action the doctor has set forth?

Or you have a client who is in recovery. He really wants to stay clean and sober because his wife is threatening to leave, and his boss has given him an ultimatum. He keeps relapsing. He wants to know why the cravings are so strong—why can't he stay clean and sober? What's going on in his head that he can't control? If you don't understand the neurobiological basis of addiction, how can you explain it to him? Is urging him to Just Say No the answer?

What about your client with PTSD (posttraumatic stress disorder), who was in combat a few years ago and is now having flashbacks, night sweats, and waves of paralyzing terror? Are you still thinking in the old way about PTSD ("it's functional") because it's caused primarily by overwhelming environmental stressors? Are you aware that the person's brain has actually changed? How might it help you to work effectively with this client if you know that the intensity or persistence of PTSD symptoms arises from neurobiological changes induced by the stressful situation? What do the neurobiological underpinnings of PTSD have to do with the types of interventions you recommend?

Here's one more example. You are assigned a client with a diagnosis of borderline personality disorder. She has a reputation among your colleagues for being impulsive and prone to suicidal gestures such as cutting and overdosing. She's been involved with your agency for a long time. They've assigned her to you because she's burned out the other workers. What is going on with this person? Do you understand her? What role does neurobiology play in her borderline characteristics? How are you going to help her? What reason do you have for thinking your interventions may be successful?

Neurobiology is still a long way from having all the answers. In fact, it's only just made its entrance. But it offers explanations that can help you answer some of these questions better than you could without it. This book suggests a few of these explanations, lays the groundwork for you to seek out many more on your own, and gives you some of the scientific background you'll need to help you understand the science-based material that you find.

3
Domains of Biological Influence on Psychological Functions[1]

This introduction to biopsychosocial knowledge would not be complete without information on the domains of biological influence on psychological phenomena. Table 1 (pp 24-25) presents biological functions that may act in concert with non-biological factors to influence emotions, behavior, and cognition, and mechanisms through which they wield this influence. Examples are included of specific conditions or symptoms that can arise from each domain.

4
Why Do We Know So Much More Now Than We Knew in the 1980s and 1990s?

At least four factors have contributed to rapid advances in neurobiological knowledge: (a) a technological revolution; (b) accumulating evidence about neurobiological structures and functions underlying psychological problems; (c) experiences with new medications; and (d) public education by mental health family and consumer advocacy groups

A) *The technological revolution*
- Brain imaging techniques allow us to see not only structures but also actual functions or processes in the living brain.
- Electron microscopes allow us to see and photograph minute structures in the brain such as neurons, synapses, and even molecules.
- Microelectrodes allow us to wire and measure the activity of a single neuron (nerve cell).

Brain imaging has brought about unprecedented breakthroughs in our knowledge. These techniques measure brain characteristics using computers to see "slices" of living brains. This information is transformed into visual images on monitors or into printouts of numerical values. Examples of techniques that show structures are CT (computed tomography) and MRI (magnetic resonance imaging). CT scans are cheaper and can detect calcified lesions better than MRIs. MRIs give a clearer picture (have superior resolution), can distinguish subtle differences in brain tissue, and can represent thinner slices of brain tissue.

Examples of techniques that show functions or processes are PET (positron emission tomography), SPECT (single photon emission computed tomography), and fMRI (functional magnetic resonance imaging). Glucose compounds, incorporating radioactive substances acting as markers that PET and SPECT can read, are injected into the blood, allowing PET and SPECT to track glucose concentrations in different brain regions. In PET scans (see Figure 1, p 27), areas of high activity (increased glucose metabolism) are represented by reds, oranges, and whites. Areas of low activity (decreased glucose

Table 1

DOMAIN[1]	MECHANISMS OF INFLUENCE	EXAMPLES OF CONDITIONS SOMETIMES RELATED TO DOMAIN
Genetic	genetic programming	bipolar disorder, schizophrenia, obsessive-compulsive disorder, Down's syndrome , major depression , attention-deficit hyperactivity disorder , tic disorders , Huntington's chorea , panic attack , dementia, autism, dyslexia, alcoholism , borderline personality , anxiety disorders
Endocrinological	hormone imbalance	some mood disorders, post partum psychosis, extreme aggressivity (when due to excess testosterone), premenstrual syndrome
Allergic/ Immunologic	autoimmune reactions antigen/antibody reactions	attention-deficit hyperactivity, depression
Metabolic	biochemical processing burning for energy (diabetes, hypoglycemia)	depression, anxiety, dementia, retardation, autism, schizophrenia

Table 1, continued

Infectious	viral or bacterial disease, e.g. encephalitis, rubella	attention-deficit hyperactivity
Traumatic	head injury oxygen deprivation damaging or destroying neurons	attention-deficit hyperactivity, impaired social judgment, impulsivity, aggressivity
Nutritional	lack of needed nutrients (proteins, vitamins, minerals)	impaired cognitive functions, apathy
Neoplastic	new tissue, as in cancer/tumor , pressure from tumors in brain	dementia, aggressivity, depression
Toxic[2]	lead, mercury, street drugs, prescribed meds, workplace chemicals	retardation, attention-deficit hyperactivity, dementia, aggressivity, learning disability, autism

1 Separation into domains is for purpose of clarity only. Domains are highly interactive with each other.
2 Behavioral and cognitive problems are increasingly being related to proliferating environmental toxins.

metabolism) are represented by greens and blues. SPECT scans are similar to PET and cheaper, but they provide less clear pictures (poorer image resolution). Functional MRIs give a series of rapid MRI images that can measure regional oxygen levels in blood vessels without injecting anything, so that no radioactive substance is administered. In addition, fMRIs can show more detail than PET about the activity of a particular region. For these reasons, fMRIs are often the imaging of choice.

In the example in Figure 1, the person with ADHD has more areas of greens and blues, indicating underactivity in certain areas of the brain where less glucose is being burned up. The person without ADHD has more areas of reds, oranges, and whites, where there's more glucose being burned up. In typical persons, the brain systems that inhibit behaviors balance the effects of systems that excite behaviors. The scan shows underactivity in more brain regions of the person with ADHD than in the brain without ADHD. But people with ADHD are often overactive.

So why should that be?

In specific brain regions, some inhibitory forces that are supposed to balance excitatory forces appear to be underactive, allowing some excitatory forces to get revved up. This in turn is associated with behavioral disinhibition or overexcitation. I'll explain more about excitatory and inhibitory actions in the brain in Part II.[2]

B) *There is new evidence about neurobiological structures and functions that underlie psychiatric illness.* Scanning technology advances have allowed us to see that psychiatric disorders involve certain specific brain structures or functions that differ from those of typical brains. Anatomical (structural) and physiological (functional, process) characteristics of the brains of people with psychiatric disorders are phenomena that we can't see without technology. These invisible phenomena relate directly to the psychiatric symptoms that we do see in daily life.

figure 1

PET Scans of Adults without and with Attention-Deficit Hyperactivity Disorder (ADHD)

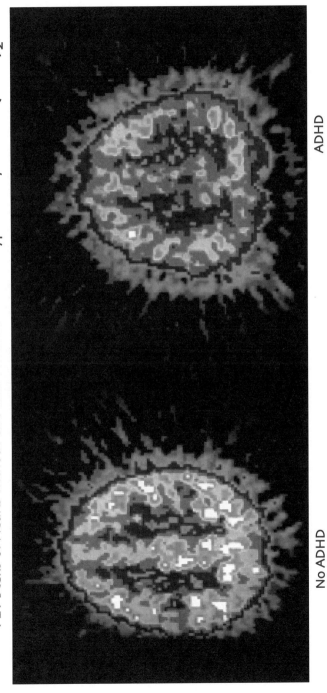

No ADHD

ADHD

Zametkin AJ, Nordahl TE, Gross M, King AC, Semple WE, Rumsey J, Hamburger S and Cohen RM (1990). Cerebral glucose metabolism in adults with hyperactivity of childhood onset. *NEJM* 323(20):1361-1366.

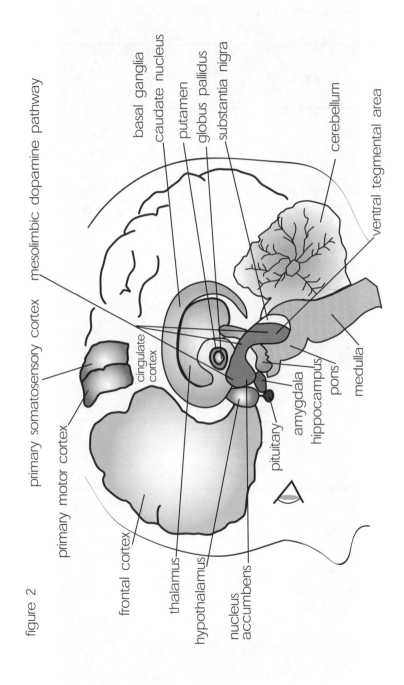

figure 2

basal ganglia
caudate nucleus
putamen
globus pallidus
substantia nigra

cerebellum

ventral tegmental area

mesolimbic dopamine pathway

primary somatosensory cortex

primary motor cortex

cingulate cortex

frontal cortex

thalamus

hypothalamus

nucleus accumbens

pituitary

amygdala

hippocampus

pons

medulla

C) *Experiences with new medications* have shown that specific symptoms—such as hallucinations, delusions, hyperactivity, compulsive rituals, or paralyzing fear in the absence of a real threat—respond to specific medications associated with each type of symptom.

D) *Mental health family and consumer advocacy groups are educating the public.* The work of these groups has increased public awareness of brain/behavior relationships and the role of neurobiology in mental illness and mental health. Examples of national groups with local affiliates around the country are NAMI (National Alliance for the Mentally Ill), FFCMH (Federation of Families for Children's Mental Health), CHADD (Children with Attention Deficit Disorder), and LDA (Learning Disabilities Association). These groups have collaborated with professionals in agencies, with the National Institute of Mental Health, and with national educational organizations to disseminate information about neurobiological disorders. These groups have also lobbied for legislation, such as parity in third-party reimbursement.

2. This example shows only two single cases, one person with and another without ADHD. In relation to this particular study, other safeguards for validity were in place that supported the conclusions. Here, we see only an illustration of a difference between the brains of two individuals who do and do not meet criteria for the ADHD diagnosis. If we are to have any confidence that the difference between the two brain images has anything to do with ADHD, we need corroborating data from multiple sources such as individual and family history, response to previous treatment, reports of scans of the brains of several people from each of the two groups (those with and without ADHD), and comparisons of medical and demographic variables.

5
The Biopsychosocial Perspective: Theoretical Frameworks

New information about brain/psyche/environment interactions is encompassed by several intellectual frameworks. In each of the following conceptual approaches, neurobiological knowledge can be an important component.

General Systems Theory (GST) represents the world as a giant living system composed of interacting and mutually influencing components (systems, suprasystems, and subsystems) that range in size from very small particles (electrons, atoms, molecules) to very large national and international forces (economic, political, cultural). The arrangement of these components is hierarchical. A system is defined as an entity with more than a single part, separated from other systems by physical or metaphorical boundaries, that receives inputs and dispatches outputs. A system exchanges inputs and outputs with other systems outside its own boundaries to carry out functions necessary for the system's survival (Bertalanffy, 1956; Miller, 1978).

Almost all systems contain smaller component systems within themselves, called "subsystems." The system itself is also a part of larger systems, called "suprasystems." Other systems on the same level in the hierarchy with a system are called "parallel systems."

According to general systems theory, systems can be living or nonliving. For example, a thermostat, a single computer, and a network of computers are examples of *nonliving systems* at different levels of complexity. An amoeba, a human being, a support group for would-be dieters, a family business, an elementary school, the United States Senate, a multinational corporation, and the International Monetary Fund are examples of *living systems* at different levels of complexity.

The thermostat (nonliving system) and the amoeba (living

system) represent very simple systems that may have few or no subsystems embedded in them. Our other examples of systems act both as subsystems to larger systems (a single computer is a subsystem to a network of computers, a person is a subsystem to the weight-control support group she belongs to) and as suprasystems to their smaller component systems (the single computer is a suprasystem to its built-in modem, the person is a suprasystem to her circulatory and reproductive systems). Any system that includes living systems within it is defined as a living system, hence organizations and government bodies are living systems. For example, the United States Senate is a suprasystem to the Senate Finance Committee, and the Senate Finance Committee is a suprasystem to each of its members.

James Grier Miller (1978), in a monumental 1,000+ page exposition, showed how general systems theory applies to human systems of all sizes. Warren, Franklin, and Streeter (1998) explain how concepts from two contemporary theories (chaos theory, complexity theory) can advance systems theory as a foundation for social work practice by strengthening non-linear analysis of human behavior in its environmental contexts.

Contemporary Developmental Psychology, as conceived by some of its leading proponents, incorporates reciprocally influencing biological, interpersonal, cultural, organizational, economic, and political variables into an understanding of human development (Kagan, 1994). These developmental psychologists expand the general systems theory constructs of interlocking, interacting systems by emphasizing the time dimension, that is, systems interacting through time. Jerome Kagan's work exemplifies a biopsychosocial perspective that meaningfully incorporates research on child and adult development through time. Rather than progression through universal stages from birth to death, development is viewed as ever-changing and highly variable between individuals, depending on their temperaments and social contexts. Development is highly responsive to "winds of change," and is seldom predictable by the kinds of demographic and interpersonal variables we typically observe and measure (Kagan and Zentner, 1996).

Epidemiology (the knowledge base for the field of public health) views the human organism in its interaction with the environment to assess risk factors and protective factors for human conditions, problems, and disorders. The concept of risk and protective factors pertains to biological and nonbiological variables in individuals and in environments, interacting through time, to prevent or minimize (protective factors) or cause or exacerbate (risk factors) problems of health/mental health and social problems (World Health Organization, 1992). Little stretching is required to meld Kagan's (1994) developmental model with epidemiological research, as these two approaches are fundamentally similar to each other and to Strohman's epigenetic model (2003), discussed later (Chapter 7). Risk and protective factors are variables that induce changes through time (for better or worse). Epidemiology emphasizes quantitative tools for data collection and analysis to generate knowledge and predict the likelihood of future events.

Ecological Theory considers mutually influencing interactions between people (human organisms) and their environments (Germain, 1991; Mattaini, 1999). In its social work applications, ecological theory has stressed the concept of environments as contexts for psychological functions. It has helped generations of social work students appreciate the interdependence between individuals, families, and their social and physical contexts. Empirical studies of ecological theory by other disciplines, such as epidemiology and cellular biology, have supplied research data that support the validity of the underlying concepts of ecological theory and theories of living systems (World Health Organization, 1992; Strohman, 2003).

Strengths Perspective. The strengths perspective is qualitatively different from the other approaches we've considered, in that it is primarily a philosophy or way of interpreting information about bio-, psycho-, and social factors rather than a theory that

32

attempts to describe and explain the world. According to Saleebey, the principles of strengths-based approaches are: (1) Every individual, group, family, and community has strengths. (2) Trauma, abuse, illness, and struggle may be injurious, but also may be sources of growth. (3) We must stop assuming that we know the upper limits of people's capacities to grow and change. (4) Collaboration is a more helpful posture than a posture of professional expertise. (5) Environments are full of resources if only we take the time to find and exploit them. (6) Caretaking is a natural and essential condition for human well-being. There is nothing wrong with wanting to be taken care of and accepting caretaking, the social norms of independence and self-sufficiency notwithstanding. All people who need care must get it. "Dependence," once equated with psychological deficit, is viewed from the strengths perspective as a natural human need (Saleebey, 2001, pp 13-18).

Strengths-based social work practice and ecological practice are consistent with a biopsychosocial perpective, assuming they are based on multifactor assessment that includes the 'bio' component. General systems theory, research-based developmental psychology, and epidemiology are inherently consistent with a biopsychosocial perspective since they explicitly include knowledge from neuroscience.

6
The Biopsychosocial Perspective: Unifying Themes

Some unifying themes of the biopsychosocial perspective appear frequently in the theoretical frameworks we've just summarized.

A. *Human emotions, behaviors, and thoughts are biopsychoso-*

ciocultural events. That is, they are responsive to complex multifaceted biological, psychological, social, and cultural environments through time. "Environment" comprises not only interpersonal milieux, such as family and peers, but also economic, political, cultural, legal, organizational, and biological forces. Biological environments supply both risk factors, such as chemical toxins or viruses in the air, water, or food supply, and drugs of abuse in the neighborhood, and protective factors, such as fruits and vegetables and therapeutic medicines.

B. *Human development is constantly in flux, in a continuous, recursive process.* Personality is not a stone statue. It is not fixed at birth or age 3, or age 10 or 21, or even 50. It is more like a lump of cold dough, solid but not frozen, changing shape as the world presses on it at one point or another. It can change shape at any time, depending on what part of it is pressed, how heavy the pressure, and how malleable it may have become from the warm hands of its modelers. The brain is plastic, i.e. amenable to change, and can produce changing emotions, thoughts, and behaviors.

C. *The once popular separation of the physical (body) from the psychological (mind) is a false dichotomy.* Psychological events always represent body/environment interactions. Emotions, behaviors, and cognitions are not due to either nature or nurture; they are responses to complex ongoing interactions between nature and nurture.

D. *Feelings, behaviors, and thoughts are situation-specific.* That is, a person responds to what he or she perceives as the requirements of the situation at that moment in time, influenced not just by his or her "personality" but also by the real attributes of the present situation. One result of this process is that when we believe we can predict people's responses based on some salient personality characteristic (e.g. Eve is an "anxious" person, Jason is a "hostile" person, or MacKenzie is a "calm" person), we are incorrect. Eve isn't anxious in every situation, only when she's in danger of being late for an appointment or when she has to speak in public. Jason isn't always a hostile person, only when someone cuts in front of him in line or calls him a wimp. MacKenzie's calm all right when she's chatting with her friends, but not when she loses her glasses or misses the bus to work.

E. *The ways we think, learn, interpret, and feel are influenced by our temperaments* (inborn, biological characteristics) interacting with learned information and specific characteristics of the present situation.

F. *In any instance of psychological function, the relative contributions of biological and nonbiological variables to the interactive process may vary.* Influence on some psychic functions may be predominantly biological, with contributions from non-biological environmental factors (as in brain disorders such as schizophrenia, bipolar disorder, attention-deficit hyperactivity disorder, obsessive-compulsive disorder, learning disability, pervasive developmental disorder, Tourette syndrome, panic attack, and some instances of major depression). Influence on some psychic functions may come predominantly from nonbiological environmental forces, with contributions from biology; posttraumatic stress disorder is the most salient example. Finally, there may be major contributions both from biological and from nonbiological environmental factors in the creation of a problem, as in addiction, some kinds of violence, and conduct disorder. No matter which influence predominates, biology and environment are involved in virtually all psychic phenomena. *Nature vs. nurture, biology vs. environment, physical vs. psychological—all these are false dichotomies!*

G. *Social work practice needs science-based knowledge.* A final unifying theme of this book is its search for strategies to integrate findings of science into our thinking about practice. We dismiss or remain ignorant of the findings of scientific inquiry—findings derived from social psychology, child development, public health, neuroscience, and other fields—at the peril of our clients. This book explores ways to enhance our ability to help clients by incorporating information from one branch of science, neurobiology, into social work assessment, problem formulation, and intervention.

7
The Biopsychosocial Perspective: Complex Adaptive Systems, Genetics, and Epigenetics

The dichotomous perspective has framed the "causes" issue as nature versus nurture rather than nature/nurture interaction. That is, it has posed the question "what percentage of condition X is due to biology, what percentage to environment?" By contrast, the biopsychosocial frame asks "how does biology interact with environment over time to arrive at condition X?"

Two errors have sometimes accompanied the dichotomous perspective. First, the word "genetics" has been substituted for "biology." The question now reads "what percentage of condition X is due to genetics, what percentage to environment?" Second, the dichotomous perspective has incorrectly assumed that genes are always deterministic. That is, once you have a gene for trait X, Y, or Z, the trait will appear. These two misconceptions diminish the importance of the "bio" as an ongoing, interactive force that is continually modified by environmental inputs, and also ignore the multiple roles played in emotions, behavior, and thought by nongenetic biological variables.

The apparent persistence of these beliefs among social work educators was addressed in a lead article in the *Journal of Social Work Education* by cell biologist Richard C Strohman (2003; Gambrill, 2003). Strohman points out that a simple linear progression from single gene to specific disease accounts for only about 2% of diseases, including sickle cell anemia and Duchenne muscular dystrophy. The remaining 98% of human diseases involve multiple genes and/or continuous interactive processes through time (Strohman, 2003).

Information from other disciplines has always been valued by some social work scholars (see for example Gambrill, 2003; Kendall, 1950). However, those of us who have long endorsed the concept of a nature/nurture interactional process have lacked the biological knowledge that could explain the mechanisms by which environmental forces impact human psychological functions. Cell biology has offered social work a scientific model that can explain and validate social work's long-held conviction about the primacy of environment in all aspects of human health, body organs, and body systems, including the brain and the central nervous system.

Current knowledge has dispelled the myth that psychology is separate from biology. Research during the decades of the 1980s and 1990s has shown that neurobiological events take place in conjunction with all psychological events, without exception. Every emotion, behavior, and cognition that we can observe is the manifestation of invisible neurobiological events. This parallel process (visible affects and behaviors, invisible actions in the brain) takes place in the context of continuous transactions between the human organism and environmental variables (both biological and nonbiological). Environmental inputs strongly influence the ways the brain functions, evoking changes in chemistry and levels of activity in different brain structures and regions. The brain's responses to the environment's inputs (*emotions, thoughts*) lead to *behaviors* that in turn impact the environment, a process encompassed by the concept of "epigenetic" (see for example Egger, Liang, Aparicio et al, 2004; Gould and Manji, 2004; Fazzari and Greally, 2004; Jablonka, 2004).

The epigenetic model includes three pathways in the body: the genetic, which is familiar to most of us; the polygenetic; and the epigenetic (not to be confused with Erikson's epigenetic model, which is not biological) (Schriver, 1995). The recent influence of epigenetic analysis is suggested by its prominence in scientific journals. Our PubMed search using the single term "epigenetic" yielded 4,151 hits, the majority of them published since 2000. Most abstracts pertained to the role of nongenetic mechanisms that influence and alter the process of gene expression. *Social Work Abstracts*, on the other hand, listed only seven citations for "epigenetic," with only two of these published since 1987 (Chatterjee, Bailey, and Aronoff, 2001; Fish, 2000).

The *genetic* pathway is the route by which a specific single gene influences a *phenotype*, defined as "what the organism looks and behaves like" (Strohman, 2003, p 190). The process is initiated within cells as transcription, the copying of genetic information from the DNA molecule (deoxyribonucleic acid) to particular forms of RNA molecules (ribonucleic acid). RNA then transfers the information to sequences of protein molecules that form the structures of cells.

The *polygenetic* pathway involves several genes acting along pathways from the level of genotype to the level of phenotype. The most serious premature killers in the United States (cancer and cardiovascular disease), as well as most psychiatric disorders and addicted states, are polygenetic, i.e. subject to influence from multiple genes.

However, the polygenetic model cannot explain phenotypal outcomes unless it is understood to operate through the third pathway, the *epigenetic*. Diagram 1 shows how the epigenetic model incorporates inputs from the environment by inserting intermediary steps between genetically coded proteins and phenotypic structures and functions (Strohman, 2003). The epigenetic model emphasizes the role of networks that influence the ways genes are "expressed" (manifested in functions). These networks, which receive, incorporate, and process inputs from the environment, are referred to as complex adaptive systems. Cellular structures, entire cells, and organs operate within these larger complex systems.

How many of us believe that all inherited characteristics come from genes that we acquire from our parents? Or that the only exception to that rule is gene mutation? Those of us who

Diagram 1
Genetic and Epigenetic Regulation

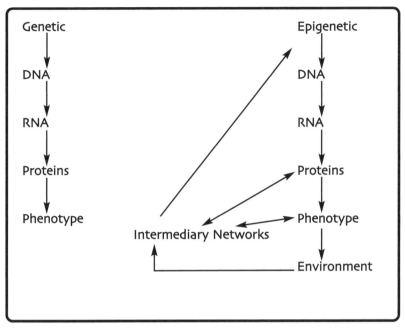

Adapted from RC Strohman (2003). Genetic determinism as a failing para-digm in biology and medicine: implications for health and wellness, JSWE 39(2) pp 175, 191. Original in MS Jamner and D Stokols (eds), *Promoting Human Wellness: Frontiers for Research, Practice and Policy*. University of California Press.

haven't taken courses in biology in recent years are often sur-prised to learn that changes can be inherited that don't come from genes at all. In complex adaptive systems, *epigenetic changes in cells are often heritable.* This is because initial cell responses not related to genes may persist over time and become stable. Change is then transmitted to daughter cells dur-ing mitosis (cell division) (Strohman, 2003, p 175).

Several layers of networks, such as gene circuits, metabolic networks, cell structures, membranes, and networks of cells, are distributed at many levels in the organism. These levels receive inputs from external environments either directly or indirectly, and dispatch outputs to external environments. Thus cells, body organs, and the human organism interact with the outside world. Gottesman, a leading expert on the genetics of schizo-phrenia, notes the applicability of a complex adaptive systems model to the etiology of schizophrenia.

> The variation observed in individual differences for normal and psychopathological behaviors has genetic factors as a major contributor at the most distal end of a complex gene-to-behavior pathway. Research into the etiologies of such major mental diseases as schizophrenia is facilitated by adopting the approach used for complex adaptive systems as pursued by those who study coronary artery disease and diabetes (Gottesman, 2001, p 867).

Two facts about genes are central to our knowledge. First, the activity of genes does not stop at conception or birth. Genetic activity can be turned on (or off) throughout the life cycle. Second, changes in gene expression (how a gene manifests itself in a phenotype) occur through changes in *cytoplasm* (the part of the cell that surrounds the nucleus). These changes can be passed on to offspring. In addition, enzymes, catalysts of many transactions at different levels, can change chromosome structures by marking sections of DNA chemically without actually changing the genetic code. What this appears to mean is that although gene codes themselves change only through mutation, the pathways through which they wield their influence are subject to other inputs beside genetic, and sometimes these other influences prevail. An estimated 20,000-30,000 different genes are present in each person. Estimates made up to three years prior to the release of results of the Human Genome Project in 2003 had been far higher, as high as 100,000 (Pennisi, 2003).

Let's consider that each of these genes may work together with other genes to produce polygenetic effects. Polygenetic actions are controlled by numerous interactive intermediary networks. As if that weren't complicated enough, *each of these intermediary networks has its own rules and procedures, and all levels are interactive with each other* (Edelman, 2004; Strohman, 2003). How can our bodies make sense out of anything so complex? How can our brains put together coherent, purposeful behaviors and thoughts? Some neuroscientists have proposals for answers to those questions, which we'll look at in Chapter 22.

It should be immediately apparent that the level of complexity for anything beyond the progression from single gene to single protein stretches the human imagination. Not only is it expecting too much of social workers to carry out assessments of these complex processes; the state of the art of neuroscience itself is far from being able to trace complex pathways from genes to phenotypes.

What neuroscience has been able to show are correlations between alterations in *specific brain regions, structures, and functions* and common human experiences, both typical (such as learning, affiliation, attachment, and aggression) and atypical (such as disturbed emotions, bizarre or antisocial behavior, addiction, childhood developmental disorders, adult personality change, and elderly dementias) (Andreasen, 2001; Carlson, 2001; Purves, Augustine, Fitzpatrick et al, 2001).

That is, neuroscience has succeeded in identifying relationships between activities in the brain and these various observable phenomena, even though it is seldom able to trace the exact pathways by which the connections are made.

The *ways* we respond to the environment are strongly influenced by genetically based characteristics, often called temperament. Children with different *temperaments* not only respond differently to their environments, but *also evoke different patterns of response from* their environments, responses that are repeated over and over. Tense, shy, or irritable children elicit a different set of responses from those evoked by outgoing, relaxed children, thus partially shaping their own environments (Kagan, 1994).

These findings in no way suggest that family environments are not important in children's emotional well-being. Most of us know from our own experience that our families were critical to our sense of security, emotional well-being, and interpretation of the world. What the findings do indicate is that biological factors are much more important than we previously thought, and operative in ways never imagined by popular developmental theorists such as Freud and Erikson. Ironically, Freud, a neurologist, believed strongly in the primacy of biology as a determinant of psychological function, but was unable to grasp the real nature of biological influences. The lack of research tools available during his lifetime to identify neurology/psychology connections led him to metaphors (e.g. ego, id, superego) associated inaccurately with early biological developmental events (e.g. oral, anal "fixations"). Were he alive today, he would have seen many of his ideas discredited, but neuroscience has validated his early conviction about the importance of biology in mental function.

8
Research Findings: Adding Neuroscience to the Biopsychosocial Potpourri

One of the premises of this book is that research-based knowledge is essential for social work practice—hardly a novel point of view. This belief is a central tenet of both the NASW Code of Ethics and the Council on Social Work Education Educational and Policy Accreditation Standards. Yet studies suggest that research and practice are still in separate worlds, in which research findings that can inform practice and policy go unused (Scott, 2002; Biegel, Johnsen, and Shafran, 2001).

Research findings are the foundation of this book. They are used to provide practice-relevant information about neurobiological aspects of individual behavior, emotions, and cognitions that take place in everyday ordinary life in typical child and adult development, atypical (psychiatric) conditions, substance abuse, and addiction. Research findings pertaining to biological variables are drawn from the disciplines of neuroanatomy and neurophysiology, psychiatry, developmental psychology, genetics, molecular biology, epidemiology, pharmacology, and radiology. Since the subject of the book is neuroscience for social work practice, we draw mostly from the natural and medical sciences.

Yet integrated biopsychosocial knowledge requires major reliance on social sciences such as social psychology, anthropology, sociology, economics, political science, social work, and family studies. I advocate an integrated biopsychosocial perspective—so am I not guilty of neglecting the psychosocial forces that shape the lives of our service consumers, in the same way that I criticize social work for having neglected the biological?

Critics could certainly argue this position if it were possible to incorporate social research extensively and still keep the book to a manageable size—but it's not. (In some of my other writings, I have had to make the converse choice of focusing almost exclusively on environmental nonbiological variables.) The sheer volume of research in all these disciplines makes it impossible to present the social science research findings beyond illustrative material. Literally hundreds of thousands of books have been published spelling out social aspects of the human psyche and the human condition. So our attempt to integrate research from all of the areas mentioned above, not just those that directly address neurobiology of psychological phenomena, is limited for the most part to illustrative examples taken from real situations that social workers deal with every day. Practice examples in the book are developed in the context of biopsychosocial assessment. I try to show, for each individual and each family, the ways that social forces interact with neurobiological aspects. In the process of this analysis, I make occasional references to findings from social research. Although my use of research from the natural sciences is also minute in light of the vastness of the total volume of that body of research, I try to review some of its highlights in a systematic fashion. No such attempt is made with respect to the social sciences.

This book is based on the conviction that social work practice—whether we designate it "clinical," "generalist," or by field of practice—needs interdisciplinary knowledge to maximize the help we can give to clients. In this volume I try to add some knowledge about neurobiology to the already enormous social work knowledge base pertaining to the psychosocial.

If the entire range of disciplines were covered equally, integrated biopsychosocial research findings should make extensive use of both qualitative and quantitative research. Both are important and useful under certain conditions. They perform complementary functions, so that combining both approaches allows social work to incorporate the advantages of both. It has long been understood that unlike qualitative research, well-constructed quantitative research allows us to generalize, whereas well-constructed qualitative research allows us to probe specific situations, respondent experiences, and respondent feelings in ways that quantitative studies of populations cannot. An example of our own combined approach illustrates the usefulness of both.

We conducted a set of studies of views of parents of children with psychiatric disabilities about professionals, and of professionals' beliefs about parents (Cournoyer and Johnson, 1991; Johnson and Renaud, 1997). Qualitative approaches enabled us to learn in depth about the beliefs and experiences of individual parents, and to understand the complex situations that engendered their views. For example, some parents told us about their expectations for their newborn child; the ways in which the developing child fell short due to the disability; how the child's suffering was prolonged and intense; how parents coped with their own grief, overpowering at times, arising from the child's pain; how they feared for the child's future and experienced chronic worry about the limits of their abilities to protect the child from failure and rejection; how exhausted and discouraged they often felt from demands of daily caretaking; and the cascade of emotions that professional behaviors elicited.

Qualitative methods (using interviews) produced many illustrations of actual behaviors toward parents that practitioners engaged in, sometimes with negative impact, sometimes with positive. The contexts of these professional behaviors were reported by parents so that we could understand the situations in which these behaviors were likely to emerge. We obtained a wealth of examples of different specific behaviors—what worker A, B, or C said and did—together with descriptions of practice situations that brought forth the behaviors. Those insights could not have been gained by quantitative methods alone.

We used the themes uncovered in qualitative in-depth interviews together with existing published qualitative research to choose variables for study of the views of large numbers of parents. The quantitative approach allowed us to draw conclusions about the relative importance and ubiquity of specific views among populations of parents. For example, the results showed that large numbers of parents felt they had been blamed for their child's disability by professionals. Fewer parents reported that workers engaged in other types of unethical professional behavior, such as questionable fee charging or failure to observe appropriate client/worker boundaries. Most parents reported that workers were courteous, even some of those workers perceived by parents to hold attitudes of blame or devaluation of the parents. The negatives were often subtle in that they did not take place in a context of overt discourtesy. That information helped us recommend changes in certain practice behaviors that could diminish negative and increase helpful interventions.

In contrast to our study that utilized a combination of qualitative and quantitative approaches, almost all research reported in this book has been generated by quantitative (numerical) methods. That is because the kinds of data related to roles of neurobiology in psychological functioning are obtained in this way. Brain scans, laboratory experiments, epidemiological studies of populations, tests of safety and efficacy of different interventions, blood studies, genetic studies, and correlations between any of these areas—all rely on quantitative methods. Our choices of types of research are thus directed by the subject matter.

Among this universe of quantitative studies, how did we choose specific pieces of information to include or exclude? Here, we acknowledge that our choices have had to be subjective—there are no guidelines or published research on the degree of relevance of any study to social work practice. We have had to draw on the author's many years of practice experience, combined with the collected wisdom of countless researchers, writers, and practitioners.

The procedures we followed in choosing which databases to use also varied according to the subject matter. We consulted Social Work Abstracts in all of the areas. In addition, for child and adult development, we used major psychology, sociology, and anthropology databases. For psychiatric and substance abuse disorders, we used medicine and psychology databases as well as web sites for federal agencies such as the National Institute of Mental Health (NIMH), the National Institute on Drug Abuse (NIDA), and the National Institute on Alcohol Abuse and Alcoholism (NIAAA). For multiple routes to quality of life, we used those already mentioned and added a few general internet sources. For connections between all of these topics and macro forces—large-scale economic, political, cultural, and legal—we used databases for economics, political science, sociology, multiculturalism, and law.

It was not possible to do exhaustive literature reviews in so many areas, so we made decisions during the process of searching to exclude interesting abstracts that did not directly illuminate our areas of interest. We emphasized the following questions as guides to including or excluding abstracts. (1) Does the abstract address a topic that is typically relevant in social work practice? (2) Does the research appear in a (blind) peer-reviewed journal? (3) Does the reported research appear to conform to well-established principles of quantitative or qualitative methodology? We were also interested in what research was available on the views of service consumers or that had been carried out in part by consumers, but lack of such research was not an exclusion criterion for abstracts.

In addition to these guidelines, we referred often to works on evidence-based practice that are appearing with increasing frequency in the social work literature. These writings can be found by using the keywords "evidence based practice." From among five question types (effectiveness, prevention, risk/prognosis, assessment, and description) (Gibbs, 2003) our reviews of research most often involved *risk/prognosis* and *assessment* of etiologies, and *effectiveness* of interventions. I highly recommend the Gibbs reference for readers who want support in finding reliable and valid evidence.

9
Assessment and Intervention Planning Tyrone, ADHD, and a Family under Stress

Tyrone is a six-year-old boy referred to a child mental health clinic for treatment by a social worker in Special Services in a suburban school. The school's consulting psychiatrist has ruled out bipolar disorder and advised Special Services that Tyrone meets criteria for a diagnosis of attention-deficit hyperactivity disorder (ADHD). Tyrone jumps out of his seat, talks out of turn, picks fights with other children, does not listen, is disrespectful and rude to adults and children, yells and sometimes hits when frustrated.

He is the second of three sons of Will and Edith M. Will is

in the sales department of a large insurance company. Edith was a nurse, but is now at home full time. According to Will and Edith, Tyrone's brothers, ages eight and three, pose no special problems. Tyrone is in the first grade, reads and does arithmetic at his grade level. He is very bright (WISC Full Scale IQ is 132).

Will and Edith are African American and recently bought a small house in an ethnically mixed middle class neighborhood. Will is the elder of two children of middle class parents. Edith is one of six children of urban working class parents. Both have close ties to families of origin. Most family members are in a large city two hours away from their new home. Will's work has required several moves around the country. The longest time the family has stayed in one location has been two years. Edith thinks that Will could get a good job in a permanent location if he were more aggressive. Will is skeptical that such jobs are available.

Tyrone was the product of an uncomplicated pregnancy and an uneventful delivery. According to Will and Edith he was a difficult baby, cried much more than either of the other two children, and woke often during the night. At about one year of age he became a head banger, which Edith says upset her greatly, but this behavior ceased after a few months. Tyrone attended nursery school at age four, but the school asked his parents to withdraw him after six weeks because of his unruly behavior. Will recalls an incident in which he was outside the classroom waiting to pick Tyrone up. Tyrone came out, saw a mother of another child, ran up to her, and kicked her in the shins. Will was mortified and reports he hit Tyrone on the behind and yelled, "Tyrone, why did you do that?" Tyrone said he didn't know.

Tyrone attended kindergarten part of the next year, but then the family moved. Because the school year was almost over, he was not re-enrolled in school until the following year. The kindergarten teacher had expressed concern about Tyrone's behavior and referred him for evaluation, but because of the move it was not completed.

Edith remembers that one of her brothers as a child was a "mild version" of Tyrone. She is very close to her mother, a widow, who is ill with diabetes and hypertension. Edith had been upset by the frequent moves and at having to be far away. She had expected to see her mother often when they moved to their new home, but Will's and the children's schedules have made this difficult.

Edith and Will agree that they got along very well until Tyrone was about three and his out-of-control behavior was a constant challenge. Before that, they hoped Tyrone's behavior was just the "terrible twos" and would diminish. They acknowledge that they fight a good deal, almost always around Tyrone. Each feels the other handles him wrong. Will advocates strict discipline involving corporal punishment, but Edith feels he is too harsh. Nevertheless, she is so frustrated she also hits Tyrone, although she doesn't think it does any good. Will criticizes her for giving in to his tantrums, which Edith admits she does because she is too tired to hold out indefinitely. Tyrone has tremendous energy, is still going strong by nine pm, by which time Will and Edith are exhausted. Some relatives of both Edith and Will have expressed disapproval of their handling of Tyrone ("Maybe he just needs a little more loving," or "A good swat on the rear might help, don't you think?").

The M family is under financial stress because of mortgage payments. Edith would like to go back to work but is unable to do so because of Tyrone's behavior. Babysitters seldom return after the first time, and the only after-school day care facility in the area has told the parents they cannot accept Tyrone. Edith and Will say that church is a positive force for all the family.

My assessment follows. Before you read it, I recommend that you do your own assessment, using the protocols in the workbook (see workbook Part I, Tyrone, ADHD, and a Family under Stress). After you've worked through the process, compare the two. I don't claim to have the "answers." Your ideas may differ from mine, and there is always room in assessment for different opinions.

ASSESSMENT. Data were assembled from the Tyrone vignette onto instruments using the tools in the workbook (genogram, ecomap, risk factor/protective factor matrix acronym EPICBIOL, and the expanded ecomap).

1) Relevant family history: maternal grandmother's illnesses, maternal uncle's "mild version" of Tyrone's characteristics, Tyrone's difficult infancy (crying, waking), head banging, hyperactivity, explosiveness, limited sleep.

2) Known or probable aversive exchanges.
Within the nuclear family:
Between the two parents.
Between each parent and Tyrone.
Between the family and external systems:
Between each parent and members of extended families on both sides (the vignette does not specify which family members).
Between Tyrone and peers.
Possible (no data): aversive exchanges between the parents and school personnel? Even in the absence of overt conflict, parents usually feel anxious and under scrutiny by professionals when their child is troublesome.
Possible (no data): frequent punishment of Tyrone in school because of his disruptive behaviors? A common response of overloaded teachers who must meet needs of many children at once.
Arenas for conflict:
Tyrone's clashes with teachers, peers, and parents (no information about siblings).
Parents' disagreement about handling Tyrone's behavior.
Edith's unhappiness with Will's job.
Critical remarks by members of extended families about Edith's and Will's child-rearing practices with Tyrone.

3) Stresses, problems, and needs underlying the aversive exchanges (risk factors)

Disruptive and aggressive behaviors by Tyrone.

Disruption arising from frequent moves due to Will's job.

Separation of nuclear family from supports.

Attributions of Tyrone's bad behavior to parental deficits (by the parents themselves, extended family members, and probably some of the school personnel).

Lack of child care due to Tyrone's behavior, leading to:

Lack of respite for Edith, loss of work outside the home.

Financial stresses due to need for Edith to be available for Tyrone.

Family life is mostly punishing for parents because of constant management needs of Tyrone (not much opportunity for family to enjoy time together).

Little opportunity for parents to have fun as a couple.

Lack of services (so far) provided by the school to meet Tyrone's special needs.

4) Strengths, resources, assets (protective factors)

Parents well educated, have marketable skills.

Will appears to have a good-paying job.

Parents report getting along very well before Tyrone's difficulties surfaced.

Apparent lack of difficulties with the two other children.

Closeness to extended families.

Parents' apparent commitment to helping Tyrone.

Tyrone's condition probably treatable.

Sufficient family resources to get help for Tyrone when specific needs for help are identified.

Research on conditions like Tyrone's shows effective interventions are available (see discussion of databases, Chapter 8).

Support to parents from the church.

Support and advocacy groups for parents available in the community (middle class suburban).

Probably strong parental bonds with Tyrone, not obvious in vignette due to focus on problems.

FAMILY DYNAMICS

I began by reviewing a summary of research on etiology and treatment of ADHD, then used tools in the workbook to do a risk factor/protective factor analysis and an expanded ecomap showing chains of effects that emanate from the disability itself. The tendency among practitioners is sometimes to note areas of conflict and stress (which are obvious in the vignette), then attribute Tyrone's problems to parental discord and inept parenting. The expanded ecomap corrects for this tendency by highlighting the role of the disability in *generating* conflict and stress for the child and his environment.

Current research emphasizes that impulsive disinhibited behaviors of ADHD, such as interrupting, picking fights, noncompliance with adult requests, and disregard for the rights of others, arise from a biological dysfunction in specific regions of the brain. These behaviors then produce strong psychosocial repercussions. Our assessment is that a negative feedback loop has been set off and perpetuated by Tyrone's disability, rather than by bad parenting or parental disagreement. There is no evidence that the parents' conflicts pertain to all the children, nor that there was a marital problem prior to Tyrone. It is typical for children like Tyrone to generate conflict, frustration, anger, and finger pointing, because no parental interventions are successful in changing the behavior. Nothing works. Therefore parents blame themselves and each other for Tyrone's behavior and so do other people in the absence of education about the real nature of his disability.

Professionals unfamiliar with the research often reinforce misconceptions held by the parents and the extended family. For example, they may respond to failure of psychosocial interventions to diminish the troublesome behaviors by recommending more intensive applications of the same psychosocial interventions that have not worked. This approach is dictated by the belief that Tyrone's behaviors are responses to environmental inputs, not to an underlying and severe neurobiological disorder. Were Tyrone's behaviors less severe, behavior modification training for parents and teachers might be sufficient.

However, research indicates that for this degree of severity (major difficulties in social and behavioral functioning across settings), behavioral treatment alone is seldom adequate.

POSSIBLE INTERVENTIONS

1) Worker self-education. The professional begins by educating herself or himself through searches of PubMed. Minimum keywords: attention-deficit hyperactivity disorder, effectiveness, etiology. Additional keywords: age, gender, ethnicity. In addition, worker self-education includes getting information about resources and entitlements for children and families, and about local access routes to these resources.

2) Education of family and child, as appropriate for age and level of sophistication. First, explore with family and child (separately or together, according to worker's and family's best judgment) what each knows about the condition or problem. Misconceptions are the rule rather than the exception. Once the professional has familiarized himself or herself about current research on etiology and interventions, s/he can share this information with parents and the child.

Parents, not the child, usually make the decision about what action to take. Although the child's preferences should be factored in, it may make sense to see parents without the child to lay out the alternative intervention options, with potential benefits and risks/side effects of each. The choice of no treatment should be included as one option with the possible benefits and risks also made explicit.

Web sites of the National Institute of Mental Health and the American Academy of Child and Adolescent Psychiatry have materials for families. Parents should also be referred to scientific databases if they want this information. Most of us can learn a lot from abstracts on research about a condition, even if we don't understand all the technical information, and parents are no exception. The worker's educational role includes answering parents' questions about the information they have gleaned from different sources, and facilitating the parents' processing of the information. The worker could offer to help parents get started with a search right there in the office that they can continue at home, or in their local library.

3) *Supportive exploration of feelings of the child and of the parents* (separately or together). Parents and children often (but not always) want the opportunity to talk about their feelings with respect to the child's difficulties. A helpful approach is to normalize the feelings that they may have by saying "lots of kids [parents] feel embarrassed [ashamed, scared, etc] about getting into fights [getting sent to the principal's office, having to take meds, etc]," or whatever the specific issue may be. It's important, though, not to push parents (or children) to talk about their feelings if they don't want to. Conveying an attitude of respect for clients is critical to a positive relationship, including respecting their wish to keep their feelings private.

4) *Medication evaluation.* Explore parents' beliefs about and attitudes toward medication. Refer the child for a consultation with a child psychiatrist or other child psychiatric expert, focusing on the question of whether medication is advisable, which medication, and what likely effect it might have. Be prepared to respond to parents' questions about the dangers of stimulants for ADHD children and the likelihood of improvement with behavioral and other non-medication interventions, and to direct parents to specific sources of information, because general internet information is not subject to quality control and can be misleading, even dangerous. Families often are fearful of medication and prefer to try behavioral approaches. However, in cases as severe as Tyrone's, behavior modification in the absence of medication has seldom been found to be more than marginally helpful. However, once the issue of medication has been addressed, behavioral strategies could be useful as adjunct treatment (see Advice on managing ADHD behaviors, below).

5) *Referral to parent support group.* Try to learn from the par-

ents what they know about such groups, and whether one or both would like to try going to a meeting or talking on the phone with a parent experienced in the system. Many parents have found the support of other parents "life-saving."

6) *Advice on managing ADHD behaviors.* Characteristics that are unique to a child's disability, for example, getting upset over transitions, are part of the knowledge that a worker needs in order to be helpful to the family. This specific characteristic requires management strategies involving changing antecedent conditions. Many other more generic behaviors, such as poor listening or interrupting, are well suited to operant behavior modification approaches. The worker asks parents for details about the child's troublesome behaviors and suggests that they choose a single behavior to work on first. Behavioral treatment must usually start off with a limited goal and scope in order to make progress. Acting as a behavior management consultant requires knowledge and experience. If you've had no behavioral training, you should probably refer the family to someone who has this expertise.

7) *Advocacy for services and entitlements.* The worker shares what he or she knows about resources for the child and the family (services, financial, or other support). If families appear interested, the worker should facilitate a referral or at least provide the family with phone numbers of services.

8) *Ongoing collaboration with the family and the school.* A frequent obstacle to successful treatment is lack of follow-up after referral. Someone needs to be responsible for monitoring progress, which means being in contact with the parents and all providers of services, as well as being available to the parents by phone in order to respond to any concerns they may have. This mechanism should be set up at an initial team meeting that will include the parents, Special Services at the school, Tyrone's classroom teacher, and the clinic social worker. At this conference a written agreement is needed to spell out who will

do what, and by what dates. All participants will sign and will receive a copy of the agreement. If Tyrone is put on a trial of medication, there must be a plan in place for parents and teacher to keep records of his response on a daily basis. The length of time needed to judge whether the medication is effective, and whether the dose should be modified or a different medication substituted, will be decided upon at the initial meeting, along with contingency plans that depend on Tyrone's response. The monitoring role, often taken by a social worker, is critical to the success of the intervention plan.

EXPECTED BENEFITS FROM INTERVENTIONS

Tyrone
Lessened impulsivity, disinhibition, and aggressive behavior.
Improved listening, compliance.
More positive feedback from others (parents, teachers, peers).

Edith and Will
Less blame/shame/stigma.
Greater understanding of Tyrone.
More acceptance of themselves and each other.
Better management of Tyrone's behaviors.
Reduced level of exhaustion.

Siblings
Calmer parents.
Reduced conflict in the home.
Possibly more parental time and attention available.

POSSIBLE ADVERSE EFFECTS FROM INTERVENTIONS

Medication side effects (side effects of psychostimulants are almost never serious):
Loss of appetite
Remedy: give medication at times other than before meals.
Trouble sleeping

Remedy: avoid late afternoon/evening medication.
Appears lethargic
Remedy: reduce dose or try a different medication.

Child experiences stigma at school by having to go to school nurse for medication.
Remedies:
Single extended release dose, if effective.
Support group for children in the school setting.

Child stashes meds and sells or gives them to peers (usually not an issue with younger children).
Remedies:
Keep meds locked up and supervise administration.
"Process" with child, i.e. talk with the child about the behavior, explore child's ideas, explain why the behavior is a problem.
Set up rules and consequences for infractions if necessary.

POSSIBLE ADVERSE EFFECTS OF NO TREATMENT

Disruptive behaviors continue, rejection of child by teachers and peers is compounded.
As child grows older, increased likelihood of poor self-esteem, depression, substance abuse, antisocial behavior, dropping out of school.
Family life continues at chronic high-stress levels, parents drained both emotionally and physically so have reduced resources for postive parenting of all the children.

References Preface and Part I*

Preface to the Second Edition
American Psychiatric Association (2000). Diagnostic and Statistical Manual of Mental Disorders, 4th ed. Text Revision. Washington DC: American Psychiatric Association.
Health and Human Services (1999). Mental Health: A Report of the Surgeon General. Dept of Public Health, US Department of Health and Human Services. Washington DC.

Hyman SE (1997). Commentary, Science and Treatment (video). Produced by the National Alliance for the Mentally Ill, Arlington, VA, in conjunction with the National Institute of Mental Health, National Institutes of Health, Rockville, MD.

Chapter 1: *The Biopsychosocial Perspective: Introduction*
Aronson E (2004). The Social Animal, 9th ed. New York: Worth.
Beless DW (1999). Foreword to the First Edition, this volume.
Biggar M and Johnson HC (2004). Report card: has the "bio" found its place? Progress toward a biopsychosocial understanding of human behavior. Unpublished manuscript.
Caplan PJ and Hall-McCorquodale I (1985). Mother-blaming in major clinical journals. American Journal of Orthopsychiatry 55(3):345-353.
Council on Social Work Education (1995). Curriculum Policy Statement. Alexandria, VA.

Chapter 2: *What Does All This Information Have to Do with Social Work?*
Andrulonis PA, Glueck BC, Stroebel CG, Vogel NG, Shapiro AL, and Aldridge DM (1981). Organic brain dysfunction and the borderline syndrome. Psychiatric Clinics of North America 4(1):47-66.
Bettelheim B (1967). The Empty Fortress: Infantile Autism and the Birth of the Self. New York: Free Press.
Cantwell D (1972). Psychiatric illness in families of hyperactive children. Archives of General Psychiatry 27:414-417.
Fieve RR (1975). New developments in manic-depressive illness. In S Arieti and G Chrzanowski, eds, New Dimensions in Psychiatry: A World View. New York: Wiley.
Kety S (1979). Disorders of the human brain. Scientific American 241:202-214.
Lewis DO and Balla DA (1976). Delinquency and Psychopathology. New York: Grune and Stratton.
Ritvo ER, Freeman BJ, Ornitz EM, and Tanguay PE, eds (1976). Autism: Diagnosis, Current Research, and Management. New York: Spectrum.
Wender PH (1971). Minimal Brain Dysfunction. New York: Wiley-Interscience.

Chapter 4: *Why Do We Know So Much More Now than We Knew in the 1980s and 1990s?*
Castellanos FX, Giedd JN, Marsh WI, Hamburger MD, Vaituzis AC, Dickstein DP, Sarfatti SE, Vauss YC, Snell JW, Rajapakse JC, and Rapoport JL (1996). Quantitative brain magnetic resonance imaging in attention-deficit hyperactivity disorder. Archives of General Psychiatry 53(7):607-616.

Zametkin AJ, Nordahl TE, Gross M, King AC, Semple WE, Rumsey J, Hamburger S and Cohen RM (1990). Cerebral glucose metabolism in adults with hyperactivity of childhood onset. New England Journal of Medicine 323(20):1361-1366.

Chapter 5: *The Biopsychosocial Perspective: Theoretical Frameworks*
Bertalanffy L von (1956).General systems theory. General Systems 1:3.
Germain CB (1991). Human Behavior in the Social Environment: An Ecological View. New York: Columbia University Press.
Kagan J (1994). The Nature of the Child, 10th Anniversary Ed. New York: Basic Books.
Kagan J and Zentner M (1996) Early childhood predictors of adult psychopathology. Harvard Review of Psychiatry pp 341-350.
Mattaini M (1999). Clinical Interventions with Families. Washington DC: NASW Press
Miller JG (1978). Living Systems. New York: McGraw-Hill.
Saleebey D (2001). The Strengths Perspective in Social Work Practice, 3rd ed. Boston: Pearson, Allyn, and Bacon.
Strohman RC (2003). Genetic determinism as a failing paradigm in biology and medicine: implications for health and wellness. Journal of Social Work Education 39(2):169-191.
Warren K, Franklin C, and Streeter CL (1998). New directions in systems theory: chaos and complexity. Social Work 43(4):357-372.
World Health Organization (1992). Basic Epidemiology. Geneva, Switzerland.

Chapter 7: *The Biopsychosocial Perspective: Complex Adaptive Systems, Genetics, and Epigenetics*
Andreasen N (2001). Brave New Brain. New York: Oxford University Press.
Carlson NR (2001). Physiology of Behavior, 7th ed. Boston: Allyn and Bacon.
Chatterjee, Bailey, and Aronoff (2001). Adolescence and old age in twelve communities. Journal of Sociology and Social Welfare 28(4):121-159.
Edelman G (2004). Wider than the Sky. New Haven: Yale University Press.
Egger G, Liang G, Aparicio A, and Jones PA (2004). Epigenetics in human disease and prospects for epigenetic therapy. Nature 429(6990):457-463.
Fazzari MJ and Greally JM (2004). Epigenomics: beyond CpG islands. National Review of Genetics 5(6):446-55.
Fish LS (2000). Hierarchical relationship development: parents and children. Journal of Marital and Family Therapy 26(4):50-1-510.
Gambrill E (1997). Social Work Practice: A Critical Thinker's Guide. New York: Oxford University Press.
Gambrill E (2003). Looking back and moving on, and Editor's note. Journal of Social Work Education 39(2):164, 169.

Gottesman II (2001). Psychopathology through a life span-genetic prism. American Psychologist 11:867-78.

Gould TD and Manji HK (2004). The molecular medicine revolution and psychiatry: bridging the gap between basic neuroscience research and clinical psychiatry. Journal of Clinical Psychiatry 65(5):598-604.

Jablonka E (2004). Epigenetic epidemiology. International Journal of Epidemiology, May 27 (epub only as of June 1, 2004).

Kendall, KA (1950). Social work education in review. Social Service Review 24:296-309.

Pennisi E (2003). Gene counters struggle to get the right answer. Science 301(5636):1040-1041.

Purves D, Augustine GJ, Fitzpatrick D, Katz LC, LaMantia AS, McNamara JO, and Williams SM (2001). Neuroscience, 2nd ed. Sunderland, MA: Sinauer Associates.

Schriver J (1995). Human Behavior in the Social Environment, 2nd ed. Boston: Allyn Bacon.

Strohman RC (2003). Genetic determinism as a failing paradigm in biology and medicine: implications for health and wellness. Journal of Social Work Education 39(2):169-191.

Chapter 8: *Research Findings: Adding Neuroscience to the Biopsychosocial Potpourri*

Biegel DE, Johnsen JA, and Shafran R (2001). The Cuyahoga County Community Mental Health Research Institute: An academic public mental health research partnership. Research on Social Work Practice 11(3):390-403.

Cournoyer DE and Johnson HC (1991). Measuring parents' perceptions of mental health professionals. Research on Social Work Practice 1(4):399-415.

Gibbs L (2003). Evidence-Based Practice for the Helping Professions. Pacific Grove, CA: Thomson Brooks/Cole.

Johnson HC and Renaud EF (1997). Professional beliefs about parents of children with mental and emotional disabilities: a cross-discipline comparison. Journal of Emotional and Behavioral Disorders 5(3):149-161.

Kazdin AE (1981). Drawing valid inferences from case studies. Journal of Consulting and Clinical Psychology 49(2):183-92.

Scott D (2002). Adding meaning to measurement: The value of qualitative methods in practice research. British Journal of Social Work 32:923-930.

*References at the end of Part I have been separated by the chapters in which they appear because many were not included in the text to keep this introductory material as readable as possible. All subsequent sections of the book are referenced together at the end of each part, without separation into chapters, since all references are identified in the text itself and pose no problem for tracking down.

PART II

Fundamentals of Neuroscience: How Does the Brain Work, and How Do Drugs Affect It?

Fundamentals of Neuroscience: How Does the Brain Work, and How Do Drugs Affect It?

There's a lot we don't know yet, but we're learning more and more. Here's a little of what we do know.[1]

The human brain is unimaginably complex. Leading neuroscientists such as Nobel Laureate Gerald Edelman and Steven Hyman, recent director of the National Institute of Mental Health, have described the human brain as the most complex entity in the known universe.

Technological developments from the 1970s to the present have dramatically advanced our understanding of the ways the brain works to mediate our feelings, behaviors, and thinking. In this section we'll explore some of this new knowledge about essentials of the brain's actions in our psychic lives. Neurobiological actions of some psychotropic medications and street drugs will also be reviewed.

The brain's functions include receiving messages, making decisions, and sending out commands. The brain works around the clock. It never stops for rest. The fuel that keeps its motors running is a simple sugar called glucose. When any part of the brain becomes very active, it burns lots of glucose, using the oxygen it gets from the lungs. The amount of glucose indicates how intense the brain's activity is in each region. Different glucose levels show up on PET and SPECT scans as indicators of activity in different parts of the brain. The brain has only 3% of the body's weight but consumes 20% of the body's oxygen. That means it burns a lot of calories to do its work.

(Want a crash diet? Maybe try thinking harder?)

The brain has hundreds of pathways, structures, and functions.

We're going to consider only a small number of these that are salient for relationships between the brain and psychological functions of emotion, behavior, and cognition.

1. 1. The information in Part II will help you understand Parts III-VI.

10
Larger Brain Structures and Systems

As you read about larger brain structures and systems, you will probably find it helpful to refer to figure 2, page 28.

The cortex and the frontal cortex. The outer surface of the brain, called the cortex, has multiple functions. It is approximately 3 millimeters thick and is convoluted with many folds and grooves, so its actual surface area is about 2 1/2 square feet. A strip of the cortex called the primary sensory cortex receives information from sensory organs (for vision, hearing, touching). Another strip called the primary motor cortex sends out behavioral commands via axons projecting down to the spinal cord. These two relatively small cortical areas are connected by the rest of the cerebral cortex, the *association areas*. These areas carry out activities that come between sensory inputs and behavioral outputs, such as perceiving, learning, remembering, and planning and executing behaviors.

The frontal cortex is the thinking part of the brain (other parts of the brain also take part in some cognitive functions). In humans, the frontal cortex comprises nearly a third of the entire cortex. The frontal cortex continually receives messages from and sends messages to other brain regions, including the limbic system. The frontal cortex is the part of the brain that distinguishes us from other animals. The higher the percentage of the cortex that's in the frontal area, the better you can do complex cognitive tasks like planning and reasoning.

How do we stack up against other animals?

Species	Frontal Cortex as Percent of Entire Cortex
humans	29%
chimps	17%
dogs	7%
cats	3.5%

Does this mean dogs are twice as smart as cats? On average, yes. (But your own cat is undoubtedly way above average in intelligence, and most probably smarter than a lot of dogs.)

The limbic system is the deep inner section of the brain, known as the "primitive" brain. Animal species that predated homo sapiens had limbic systems, and species less neurologically developed than humans, such as reptiles, have them today. The limbic system occurs in species across most levels of complexity and differentiation. It does not depend on consciousness.

Limbic structures regulate drives (hunger, thirst, sex), emotions and passion (love, rage, joy, fear, sadness), arousal, and levels of attentiveness. The limbic system also mediates fear— another survival emotion. It transmits messages of pleasure or pain into memory. The limbic system plays a key role in determining what is salient enough to be remembered. Intensely positive memories, such as a drug high, can cause people with addictions to have cravings and risk relapse even after years of abstinence. Intensely negative memories are activated even years later as symptoms of posttraumatic stress disorder. These memories are put into storage by a biochemical process that results in actual changes in microscopic brain structures— changes that are invisible to the naked eye but occur in tandem with observable behaviors and affects.

Here are some actions of important limbic system structures.

The *amygdala* is a small but complicated organ that plays important roles in reactions to situations that are crucial for survival. It links cortical regions that process sensory information with nerve systems that signal the muscles to act. It notifies the individual of threats of danger, pain, or other unpleasant consequences, needs of infants for care, and the presence of food, water, salt, and potential mates or rivals. Fight/flight reactions occur in the amygdala. It integrates behavioral and hormonal components of emotions with autonomic reactions (involuntary bodily events such as heart rate, respiration, and intestinal activity). The amygdala mediates rage reactions. When a person flies into a rage, his or her amygdala is revved up.

The *amygdala* is also thought to be involved in the paths that recognize pheromones (sexually stimulating odors).

The *hippocampus* plays a critical role in learning and memory. It converts immediate (short-term) memories into long-term memories as relatively permanent biochemical and structural changes in neurons. Damage to the hippocampus can result in loss of short-term memory in animals and humans.

The *mesolimbic dopamine pathway* is a major reward circuit in our brain, also called the *pleasure pathway*. Why do we have reward circuits? For survival of the species. Reward circuits motivate animals to eat, drink, and copulate. The mesolimbic dopamine pathway extends from the ventral tegmental area (VTA) in the midbrain (or *mesencephalon*, Greek for middle brain), to a structure of the limbic system, the *nucleus accumbens* (NAc). The NAc receives messages transmitted by dopamine neurons along the pathway from the VTA and is crucial in the process of becoming addicted. We'll discuss its role later in Part V on substance abuse and addiction.

The mesolimbic dopamine pathway has a role in creating "privileged" memories—memories so powerful that they keep popping up even when you wish they wouldn't. A privileged memory can be either pleasurable or painful.

The *diencephalon*, an interior region of the brain near the limbic system and the basal ganglia, comprises the thalamus and the hypothalamus.

The *thalamus* has a complex role in information processing. It is a relay station that receives information from and transmits it to different regions of the cerebral cortex, the rest of the brain, the body, and the external environment.

The *hypothalamus* (under the thalamus) exerts controls over the autonomic nervous system and the endocrine system. It organizes behaviors relating to survival of the species, known as the four F's: feeding, fighting, fleeing, and fornicating. It connects directly with the *pituitary gland*, releasing hormones that stimulate the pituitary gland to release its hormones. Working in conjunction with the pituitary gland, the hypothalamus regulates sleep, appetite, and sexual activity. The hypothalamus and the pituitary gland are the major hormone regulators in the body. Stimulation of the hypothalamus can produce upsurges in sexual behavior. The hypothalamus and the pituitary gland connect via the circulatory system with the adrenal glands located on the kidneys, in a complex feedback loop known as the hypothalamic-pituitary-adrenal (HPA) axis. The HPA also has significant influence over the immune system.

The pituitary gland is known as the "master hormone" because it controls other endocrine glands, such as those in reproductive organs (testes and ovaries). It is only the size of a blueberry, but it is pivotal in creating and regulating hormones. However, this "master" itself is controlled by the hypothalamus. Hormones produced in the hypothalamus are packaged in vesicles and propelled down to the end of axons, where they are stored in nerve endings. A special system of blood vessels directly connects the hypothalamus with the pituitary gland, transporting hormones manufactured in the hypothalamus to the pituitary gland. These hormones in turn control the pituitary's production and release of several other hormones used all over the body, including the brain's natural opioids, the endorphins and the enkephalins (see Table 2, p 77).

The *basal ganglia* are a system of structures deep within the brain adjacent to the limbic system. These structures regulate normal and abnormal movement and are also implicated in many psychiatric disorders, for example paucity of movement in depression and abnormal grimaces in schizophrenia. The basal ganglia contain the highest concentration of dopamine D2 receptors in the brain; these receptors are the primary target of action for many antipsychotic medications. Basal ganglia structures include the *caudate nucleus* and the *putamen* (region called the *striatum*), the *globus pallidus*, and the *substantia nigra*. These structures interconnect with one another and with other brain regions. As part of their functions of regulating motor activity, the caudate nucleus, putamen, and globus pallidus regulate levels of inhibition. Disinhibition, a function that is adversely affected in disorders of impulsivity and control such as attention-deficit hyperactivity, is sometimes a result of inadequate activity by these structures.

The brainstem. Although there is some disagreement about whether to consider the diencephalon as part of the brainstem, it is agreed that the brainstem arises in the midbrain and descends through the pons and the *medulla* to connect with the spinal cord. The pons, a bulge in the brainstem,

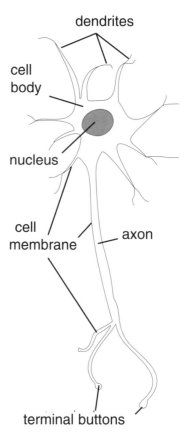

neuron   figure 3

dendrites

cell body

nucleus

cell membrane

axon

terminal buttons

is important in sleep and arousal, and in relaying information from the cerebral cortex to the cerebellum. The medulla regulates the cardiovascular system, respiration, and muscle tone.

Although some of the structures we've just discussed are small compared with the liver or the lungs, they are very large relative to the billions of microscopic structures in the brain.

11
Microscopic Structures: Neurons, Synapses, and Other Amazing Structures

Neurons (see figure 3, p 70). These tiny cells in the brain, spinal cord, and throughout the body carry out the brain's functions of receiving messages, making decisions, and sending out commands. Estimates of the total number of neurons in the human organism vary; one such estimate is 100 billion. Neurons connect with other neurons through tiny spaces, called *synapses*, that are filled with gelatinous fluid. Each neuron is thought to make from several up to hundreds or even thousands of such connections with other neurons. In the brain and spinal cord, neurons are tightly packed together in bundles. The basic structures and functions of human neurons are similar to those of less differentiated species, such as the squid, the snail, and the leech. (Did you ever think of these creatures as your first cousins?)

The neuron, like other cells in the body, is covered around its entire surface with a thin *cell membrane*. It has a cell body containing a nucleus, and many tiny filaments (threads) called *dendrites* that receive chemical messages from adjacent neurons. Each neuron has many dendrites but only one *axon*. The axon is also a threadlike structure. It sends chemical and electrical messages to the receiving neurons. (Although messages can also go from the axon of one neuron to the axon of another neuron, or from an axon to the cell body of an adjacent neuron, we will discuss mostly the more common transmission route from the axon of one neuron to dendrites of adjoining neurons.)

Neurons come in different sizes and shapes. There are as many as 200 different varieties. Some neurons have long axons (as much as a meter in length when stretching down the spine) and some have short axons. Although there is only one axon per neuron, at its end the axon divides into numerous projections with tiny bulbs at their ends. These are called *terminal buttons*. At these terminal buttons, chemical messengers called neurotransmitters are manufactured and then dispatched to other neurons.

Vesicles are round containers that store neurotransmitter molecules in the terminal buttons. Vesicles are made of the same material as cell membranes. They are manufactured in the cell body by tiny factories called *Golgi apparatus*, then transported down the axon to the terminal buttons.

Synapses (figure 4) are tiny spaces between the cell membranes of adjacent neurons. The dendrites of neurons are very near (but don't actually touch) the axons of adjacent neurons. They are separated by these tiny spaces filled with gelatinous liquid. There are an estimated 100 trillion synapses in the human body. Synapses are where the action is! Although synapses make up less than 1% of the total volume of the brain, they are the sites where the most important actions influencing emotions, thoughts, and behavior take place.

Although very important functions take place in the nucleus and the cell body, the task of transmitting messages from neuron to neuron–which happens for every activity of our psychic lives–takes place at the junctures between neurons, not in the cell body.

Presynaptic membrane. The section of the sending neuron's

figure 4

terminal button and synapse

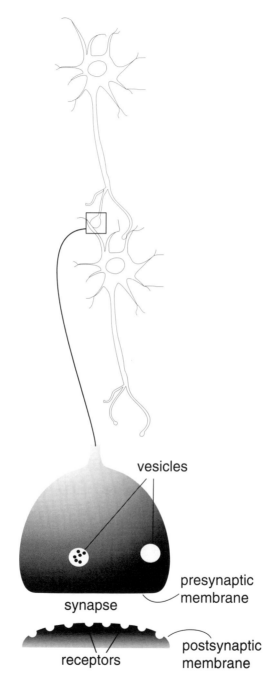

vesicles

presynaptic
membrane

synapse

receptors

postsynaptic
membrane

cell membrane (the thin membrane that covers the entire surface of the neuron) that is adjacent to the synapse.

Postsynaptic membrane. The section of the receiving neuron's cell membrane that is adjacent to the synapse.

Receptors. Protein molecules embedded in the surface of postsynaptic and presynaptic membranes. Although presynaptic as well as postsynaptic membranes have receptors, we will primarily discuss receptors on postsynaptic membranes.

Receptors have many different chemical structures. Each receptor molecule combines only with those neurotransmitter molecules that fit with its chemical structure, like a lock (the receptor molecule) and a key (the neurotransmitter molecule). The chief function of postsynaptic receptors is to alter the electrical forces (called potential) that either increase or decrease the electrical current set off in the receiving neuron. Receptors are the sites where many psychoactive drugs do their work.

12
The Brain's Natural Chemicals: Precursors, Messengers, and Enzymes

Glucose. The simple sugar that the brain uses as energy to do its work. Remember, some of the brain scans that measure how much activity is taking place in different regions of the brain do it by forming color-coded images of glucose metabolism in those regions.

Enzymes. Compounds that trigger chemical reactions without themselves being chemically changed. We have many enzymes throughout the body that perform essential functions. Enzymes can be recognized because they end in the letters -*ase*. For example, monoamine oxidase is the enzyme that causes monoamine compounds to be broken down chemically. Enzymes are synthesized in the cell bodies of neurons and transported down axons to the terminal buttons.

Precursor substances. A precursor is something that comes before something else. Precursor substances in the brain are building blocks for neurotransmitters. They are protein molecules that come from digestion of protein-containing foods. Many of these proteins have been broken down during digestion into amino acids. They are transported to the brain by blood vessels, enter the neuron, and are transformed into neurotransmitters inside the neuron.

Neurotransmitters, also referred to as transmitters, are molecules that act as chemical messengers (see Table 2 p 77). They transmit messages that direct all our psychic functions. Neurotransmitters are manufactured from precursor substances either in the axon terminal buttons (most transmitters) or in the cell body (neuropeptides). They are stored in synaptic vesicles, then are released from the vesicles into the synapse where they cross the synapse and combine with postsynaptic receptor molecules to start off a chain of events in the receiving neuron. Until recently, it was believed that neurons produced only one kind of transmitter, but we now know that many neurons contain and dispatch more than one type. Different types of vesicles within the same terminal button contain the different transmitters.

The total number of neurotransmitters is unknown, but more than 100 have been identified. Many transmitters are considered putative because it has not been possible to determine whether they meet all criteria for neurotransmitter status. According to widely accepted criteria, in order to be dubbed "neurotransmitter," a molecule must be present within a presynaptic neuron, must be released in response to an electrical current involving influx of calcium ions into the terminal button, and must have specific receptors at its target site on the post-synaptic neuron with which it can combine chemically. To carry out this function, a transmitter must be chemically constructed to fit with the chemical construction of its receptor (like a key in a lock). This process will be explained in Chapter 13 on neurotransmission.

Neuromodulators are cousins of neurotransmitters. They act like transmitters, but their action is not limited to the synaptic cleft. They diffuse much more widely through extracellular fluid (fluid outside the cell membrane).

The picture is confusing because a substance can act as a neurotransmitter in one situation, a neuromodulator in another, a hormone acting in the brain, or a hormone acting at distant sites in the body. Transmitters often have multiple functions and behave differently in different contexts. A transmitter may combine with several different receptors, all dedicated to that particular transmitter. For example, serotonin has at least 14 different receptors and changes its behavior depending on which receptor is receiving it. The neurotransmitter epinephrine in the brain is the same substance as the adrenaline your adrenal gland secretes when you get "butterflies" in the stomach. The histamine that makes you sneeze, itch, or have watery eyes is also a neurotransmitter that regulates sleep/wake cycles, hormonal secretion, and other functions. Estrogens and androgens are neuromodulators as well as those sex-related hormones that make you feel — well, you know what we mean.

For our purposes, becoming familiar with what these substances do to influence our emotions and behavior is more important than arguments about which substances fall into one or another category. Therefore, we list some of the more common neurotransmitters and neuromodulators together in Table 2, grouped in "families." Readers are referred to neuroscience texts for more detailed discussions.

Second messengers. Neurotransmitters are "first" messengers,

Table 2
Families of Neurotransmitters and Neuromodulators

Amino acids: gamma-aminobutyric acid (GABA),
   glutamate, glycine
Monoamines: dopamine, norepinephrine, epinephrine,
   serotonin, histamine
Acetylcholine
Neuropeptides
   endogenous opioids: α-endorphin, β-endorphin,
      γ-endorphin (γ-endorphin is the Greek letter
      gamma),dynorphin, leu-enkephalin, met-
      enkephalin
   hypothalamic hormones:
      corticotropin-releasing factor (CRF),
      vasopressin, oxytocin, somatostatin,
      thyrotropin-releasing hormone (TRH),
      luteinizing-hormone-releasing hormone (LHRH)
   other peptides: substance P, delta
      sleep-inducing peptide (DSIP), glucagon,
      bradykinin
Other substances that generally act as neuromodulators:
   growth hormone, prolactin, adenosine, prostaglandins,
   corticosteroids, estrogens, androgens

---

bringing signals to neurons. In order for the transmitter to act, the first-messenger signal must often be translated by the action of second messengers into a signal the receiving neuron can interpret. Second messenger molecules are activated when a neurotransmitter combines with a postsynaptic receptor, initiating a cascade of steps that allow transmission of the impulse to the receiving neuron to occur. Second messengers have other functions such as traveling to the cell nucleus and initiating biochemical changes. Examples of second-messenger molecules are calcium ions (Ca++) and cyclic adenosine monophosphate (cAMP). The actions of these and many other second messengers are detailed in neuroscience texts referenced at the end of Chapter 17.

The three most common neurotransmitters in the central

nervous system (CNS) are glutamate, GABA, and glycine. These transmitters have a general role in determining how fast receiving neurons fire. *Glutamate* (also called glutamic acid) is the principal excitatory transmitter in the brain and spinal chord. *GABA (gamma-aminobutyric acid)* is the most abundant inhibitory transmitter in the brain and is actually manufactured from (excitatory) glutamate by the action of an enzyme that removes one chemical component from the glutamate compound. That is, the most important excitatory transmitter is a close relative of the most important inhibitory transmitter—similar but not identical chemical structures, opposite actions! GABA is the transmitter in as many as one-third of the synapses in the brain. It is most often found in local circuits, where it acts through networks of short connecting neurons called *interneurons* that link many neurons in the brain. *Glycine* is a more localized inhibitory transmitter used in the spinal cord and lower brain. All other neurotransmitters transmit directives through circuits of neurons that carry out diverse specific brain functions.

GABA receptors are essential to the effects of some anxiety-reducing drugs (called anxiolytics) such as alprazolam (Xanax), lorazepam (Ativan), and diazepam (Valium) of the benzodiazepine family (popularly known as "benzos"). The type A GABA receptor ($GABA_A$) has not one but five binding sites, one for GABA itself, one for "benzos," one for barbiturates, and two others. Here's a sketch of a $GABA_A$ receptor in the membrane of a postsynaptic neuron, with these three binding sites labeled. These drugs all enhance the inhibitory effects of GABA in reducing anxiety, promoting sleep, reducing seizures, and promoting muscle relaxation.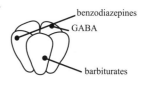

How could the brain possibly know to grow receptors for specific drugs that don't occur naturally in the brain? The brain is amazing, but is it that amazing? Has the pharmaceutical industry pulled off an extraordinary coup? Not likely. It is believed that some (unidentified) chemicals exist within the brain called *ligands*, i.e. substances that have an affinity for certain receptors. Whatever these ligands may be, presumably the brain "grew" these receptors for the ligands, not the drugs.

All muscular movement results from release of *acetylcholine* (ACh), which also facilitates dreams in REM sleep, perceptual learning in the cerebral cortex, memory function in the hippocampus, and arousal and alertness in the frontal cortex and striatum. There is no need to remember examples such as the ones given for acetylcholine. You can always look up actions of specific transmitters if you need the information, and it would be impossible (and serve no useful purpose) to try to memorize the myriad functions of different brain chemicals. We merely give a few examples to illustrate for readers how incredibly complex and how amazing an instrument the human brain is!

In the monoamine neurotransmitter family, *dopamine* (DA) produces both excitatory and inhibitory responses in post-synaptic neurons, depending on which of several different dopamine receptors is involved. Dopamine mediates movement, attention, learning, and the reinforcing effects of drugs. Cell bodies of dopaminergic neurons in the *substantia nigra* project axons to the striatum (caudate nucleus and putamen), affecting movement and inhibition. (The suffix "ergic" designates a specific neuron by the kinds of transmitters it produces, such as *dopaminergic*, *cholinergic*, and *serotonergic* neurons.)

DA cell bodies in the ventral tegmental area (VTA) project to the nucleus accumbens, amygdala, and hippocampus in the limbic system where they promote reward and reinforcement from drugs of pleasure, and to the prefrontal cortex, where they stimulate formation of memories and carry out executive functions such as planning and problem solving (see figures 6a and 6b pp 86-87, dopamine and serotonin pathways).

Almost all regions of the brain receive *norepinephrine* (NE) as input from noradrenergic neurons. "Noradrenergic" is the label used to refer to neurons that produce norepinephrine, because adrenaline and noradrenaline are synonymous with epinephrine and norepinephrine respectively, and because the alternative phrase, "norepinephrinergic," is hard to pronounce. Norepinephrine is a much more important neurotransmitter than epinephrine. Epinephrine is a minor transmitter in the brain, but it is a major hormone in the body where it is called adrenaline, sent out when needed for "fight or flight" by the adrenal gland located on the kidney.

Many of the cell bodies of noradrenergic neurons originate in the *locus coeruleus* in the brainstem, and project axons to most regions (cortex, limbic, basal ganglia, diencephalon, cerebellum). Although receptors for norepinephrine produce both excitatory and inhibitory effects, the primary behavioral effect of norepinephrine is to increase vigilance and attentitiveness to the environment, an excitatory function. Norepinephrine increases the individual's responsiveness to other excitatory transmitters.

Serotonin (also called 5-HT for its chemical name 5-hydroxytryptamine) has complex behavioral effects. It has become well known for its role in regulating mood, through the popularity of antidepressant drugs like fluoxetine (Prozac), sertraline (Zoloft), and paroxetine (Paxil), designated by their function as "selective serotonin reuptake inhibitors." These functions will be explained in Chapter 16 on actions of classes of drugs.

In addition to regulating mood, serotonin also plays important roles in control of sleep, eating, arousal, and anxiety. Drugs to change serotonin levels have been used to treat obsessive-compulsive disorder, panic attack, and nicotine addiction; prevent nausea in cancer chemotherapy; curb alcohol cravings; regulate pain; and suppress appetite in obesity. Hallucinogenic drugs such as LSD (lysergic acid diethylamide) appear to produce their effects by affecting serotonin transmission. Serotonin is able to carry out these diverse functions because it has at least 14 different receptors. The effects of the messages it transmits to receiving neurons change depending on which receptor is active.

The cell bodies of serotonergic neurons are located in clusters in the middle and lower brain regions, mostly the *raphé nuclei* ("nuclei" is plural for "nucleus") in the pons and upper brainstem. They project to the frontal cortex, thalamus, cerebellum, medulla and spinal cord, hypothalamus, amygdala, and hippocampus (see figure 6b). Again, it is not necessary to memorize this information; it can be readily accessed as needed.

These three monoamine transmitters, dopamine, norepinephrine, and serotonin have a great deal to do with psychic activity. They are extremely versatile and play multiple roles in our lives. For example, dopamine is active in the pleasures of eating, drinking, and sex. It combines with several different dopamine receptors (known as D1, D2, D3, D4, and so on). It is too abundant in schizophrenia. It is also overabundant when a person is high on many street drugs. Too much dopamine can make you feel terrific, or it can make you mentally ill. Sometimes people who come into an emergency room from overdosing on amphetamines are misdiagnosed as having an acute episode of schizophrenia because the symptoms can look a lot alike.

Serotonin and norepinephrine are believed to be released from varicosities (swellings) in branches of axons rather than from terminal buttons. However, some do so near postsynaptic membranes, so they appear to behave as they would in conventional synapses.

Histamine neurons are concentrated in the hypothalamus and are involved in transmitting signals from the hypothalamus to the cerebellum. These neurons send a small number of projections to the brain and spinal cord, mediating arousal and attention and explaining the drowsiness side effect of antihistamine drugs. Histamine is released from cells around the body in response to allergic reactions.

Neuropeptides consist of chains of 3 to 36 amino acids linked together in a mix-and-match fashion. Amino acids are molecules with different arrangements of carbon, nitrogen, hydrogen, and oxygen. Neurons manufacture their own precursors for these neuropeptides by building giant molecules called polypeptides. Enzymes break down polypeptides into smaller segments of amino acids and then reconstitute them into diverse molecules, each one some form of amino acid chain. These new amino acid molecules are peptide neuromodulators and neurotransmitters.

Many peptides known to be hormones also act as neurotransmitters. Peptides often cohabit terminal buttons with other types of transmitters, such as monoamines, and are released in conjunction with these other transmitters.

Among the neuropeptides listed in Table 2, a range of different functions are carried out. The *endogenous opioids* are thought to be agents of alternative treatments such as hypnosis, acupuncture, and yoga. The word "opioid" refers to natural (endogenous) brain chemicals, whereas the word "opiate" refers to drugs such as heroin and morphine. They got their names from the opium poppy, cultivated for 5000 years and used as an analgesic since at least the 16th century. The neural systems activated by opioids may suppress the perception of pain, inhibit defensive reactions in some species such as fleeing and hiding, and stimulate reward circuits ("pleasure pathways"). Typically these drugs multiply manyfold the similar effects of the natural opioids.

Hypothalamic hormones (hormones synthesized by neurons in the hypothalamus and dispatched to the pituitary gland) regulate secretion of hormones by the pituitary gland that have target sites all around the body. Two neurotransmitters in the group, oxytocin and vasopressin, are manufactured in the hypothalamus, then transported to and then released from the pituitary. *Oxytocin* not only initiates the birth process in human mothers and stimulates milk production in the breast, but also stimulates "affiliation"—sexual attraction, pair bonding and paternal caretaking in animals, and the urge to cuddle!

13
Neurotransmission: How the Brain Sends and Receives Messages

The brain sends messages through the firing action of neurons, a process that goes on continuously, day and night, waking and sleeping, **24/365**. The firing action of a neuron consists of sending electrical impulses through the entire neuron, and then setting off chain reactions of similar firing action in adjoining neurons.

figure 5a

neurotransmitters bind with receptors

Neural (nerve) impulses are motions of electrically charged particles called ions that pass back and forth across the neuron's membrane. The impulse travels the entire length of the neuron (starting with the dendrites and ending at the axon's terminal buttons). It is propelled down the neuron like a row of dominoes, as each ion exchange across the cell membrane sets off another ion exchange in the next section of the neuron.

These electrical impulses are then set off in adjoining neurons by chemical messengers, the neurotransmitters, that cross the synapse from neuron to neuron. This network of interconnecting neurons is so complex that, by comparison, the world's most complicated computer looks like a child's toy.

The process of neurotransmission. Several events take place in this process. (1) Neurotransmitters enter a receiving cell by crossing the synapse from sending neurons and binding with receptors on the postsynaptic membrane of dendrites of receiving neurons (see figure 5a). (2) When the neurotransmitters bind with postsynaptic receptors, an electrical impulse is set off in the receiving neuron and travels as electrical current down the neuron, from the dendrites, through the cell body and the axon, to the terminal buttons (see figure 5a). (3) Electrical current arrives at the terminal buttons (figure 5b).

(4) When this electrical message arrives at the terminal buttons, it sets off biochemical events in the terminal buttons that cause neurotransmitters to be released into the synapse (see figure 5c). These messages are carried by the neurotransmitters from the presynaptic membrane of the sending neuron, across the synapse, to the postsynaptic membrane on the dendrites of the receiving neuron.

To understand neurotransmission, it is sometimes helpful to think of the process in two phases: transmission of electrical current down the neuron, and biochemical events at the terminal buttons. (The process is actually much more complicated, but this simplified model is intended to give an overall sense of the process to readers not versed in physiological psychology.) It is important to remember that the electrical activity itself is also biochemical. Ions are created by atoms or molecules of chemical substances losing or gaining electrons, thereby becoming electrically charged.

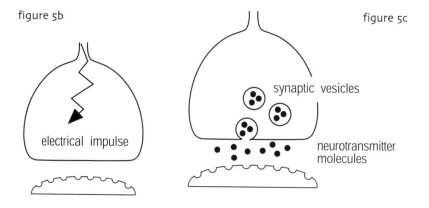

figure 5b

electrical impulse

figure 5c

synaptic vesicles

neurotransmitter molecules

These messages can be either excitatory (telling the receiving neuron to fire at a faster rate) or inhibitory (telling the receiving neuron to fire at a slower rate). The rate of responding of the receiving neuron depends on whether there is a preponderance of excitatory or a preponderance of inhibitory messages coming into the receiving neuron. Remember, the neuron may be receiving several hundred messages at the same time from other neurons adjacent to it. When the chemical messengers called neurotransmitters arrive at the postsynaptic membrane, they combine with postsynaptic receptor molecules. Each kind of postsynaptic receptor molecule only receives the specific neurotransmitter molecule whose chemical structure fits with it. That is, each different neurotransmitter molecule has its own special receptor molecule.

However, sometimes receptor molecules can be fooled. Sometimes they accept imposters, such as therapeutic or street drugs that are similar but not identical in chemical structure to the natural neurotransmitter that fits with the particular receptor. This will be explained more fully in Chapter 16.

The concentration of various neurotransmitters in the synapse is regulated by *reuptake* and *enzymatic deactivation*. In reuptake of transmitters into presynaptic terminals, other molecules, called *transporters*, carry transmitters back up from the synapse into presynaptic neurons. By removing transmitters from the synapse, the reuptake process ends one round of electrical impulse and release. Reuptake causes the overall amount of the transmitter substance in the synapse to diminish. Enzymatic deactivation consists of an enzyme setting off a chemical reaction that destroys the neurotransmitter by breaking it down into end products, called metabolites. The end result here is the same: decrease in the amount of transmitter substance available to do its work of crossing the synapse and combining with postsynaptic receptors on the surface of the receiving neuron.

Pathways for sending messages
A pathway is a bundle of interconnecting neurons that performs

specific tasks. Transmission pathways for different neurotransmitters extend throughout the brain. Examples are dopamine pathways and serotonin pathways (see figures 6a and 6b). Notice that in many areas the different transmitter systems run parallel to each other. That means they pass through and act in many of the same regions of the brain.

An example of a bundle of neurons is the mesolimbic dopamine pathway, which carries pleasure messages from the ventral tegmental area (VTA) in the midbrain to the nucleus accumbens (NAc) in the limbic system (see figure 2, p 28).

figure 6a

Dopamine Pathways

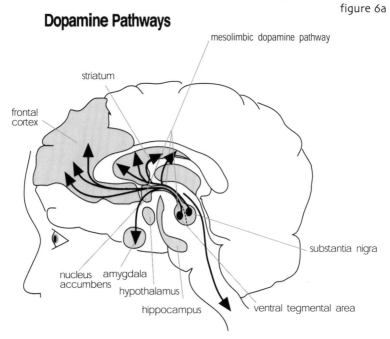

mesolimbic dopamine pathway

striatum

frontal cortex

substantia nigra

nucleus accumbens   amygdala

hypothalamus

hippocampus

ventral tegmental area

14
Neurotransmitters: Synthesis To Release in Four Steps

(1) *Synthesis of neurotransmitters* (or how your hamburger or tofu is transformed into neurotransmitters). First, you eat pro-

figure 6b
Serotonin Pathways

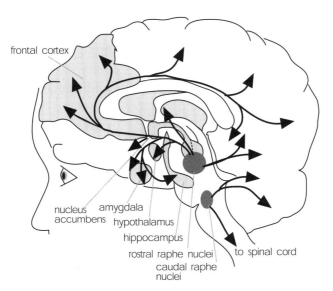

frontal cortex

nucleus accumbens amygdala
hypothalamus
hippocampus
rostral raphe nuclei to spinal cord
caudal raphe nuclei

tein-containing foods. Second, your digestive system breaks the foods down into its protein components, amino acids, which are the building blocks for neurotransmitters.

Third, your circulatory system carries these amino acids to your brain. These molecules are called precursor substances because they are precursors—they come before—neurotransmitters. The blood vessel wall acts as a filter called the *blood-brain barrier*, because it allows some but not all molecules to pass through (see figure 7a).

figure 7a   blood vessel

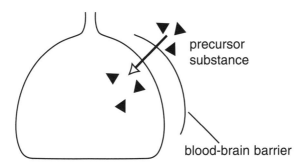

precursor substance

blood-brain barrier

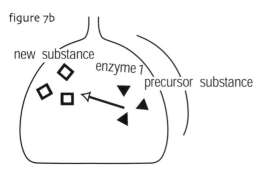

figure 7b

The precursor substances cross through the walls of blood vessels and through the cell membrane of neurons. Neurotransmitter factories are inside the neurons. Once these precursors get into the neuron, the neurotransmitter factory swings into action with the help of enzymes.

Fourth, enzymes in the terminal buttons promote chemical reactions that convert the precursor molecules into a new substance (precursor + enzyme = new substance) (see figure 7b). Let's call the first enzyme that acts on the precursor substance *enzyme 1*.

Fifth, the transformation from precursor substance to neurotransmitter molecule can involve one change (using one enzyme), or there can be a series of changes, a series of different new substances, using several different enzymes. Let's call the second enzyme in this chain of events *enzyme 2*. Enzyme 2 promotes chemical reactions that convert the new substance either into neurotransmitter molecules or into another new substance. We are using the term "new substance" as a generic term for any chemical compound that is produced during this process of transformation from precursor substance to neurotransmitter (see figure 7c). Once the final products have been manufactured (neurotransmitter or neuromodulator molecules), the life cycle of the neurotransmitter is at phase (2), storage.

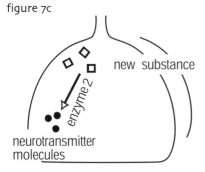

figure 7c

(2) *Storage of neurotransmitter molecules.*
To prevent various enzymes from continuing to change neurotransmitters into other compounds, the newly creat-

ed little transmitter molecules need protection, so they get into the round containers called vesicles. They are stored in the vesicles until the next phase of release (see figure 8a).

(3) *Electrical impulse.* The electrical impulse coming down the axon stimulates release of the neurotransmitters from their vesicles into the synapse. Here's how.

Electrically charged calcium particles (calcium ions) sit outside the neuron's cell membrane in the region of the presynaptic membrane (see figure 8b). The electrical impulse comes down the neuron (starting from the dendrites, through the cell body, and on down to the terminal button via the axon).

(4) *Release.* When the electrical message hits a terminal button, the electrical current stimulates channels in the cell

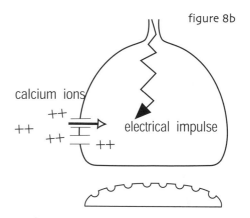

figure 8a

synaptic vesicles

neurotransmitter molecules

figure 8b

calcium ions

++
++
++   ++

electrical impulse

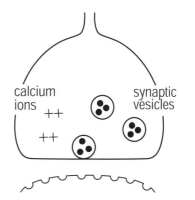

figure 8c

calcium ions   synaptic vesicles

++
++

figure 8d

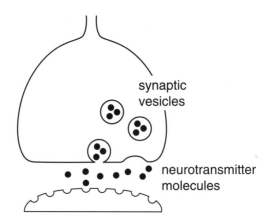

membrane to open and let the calcium ions into the neuron (figure 8b). There, the calcium ions in turn stimulate the vesicles to move to the presynaptic membrane. The vesicles fuse with the presynaptic membrane, open up, and spill their contents (the neurotransmitter molecules) into the synapse (see figures 8c, 8d).

After dumping their contents into the synapse, the vesicles cease to exist. The membranes of vesicles are made of the same material as the cell membrane itself, so as they fuse, they become part of the cell membrane. In order to replace lost vesicles, chunks of the terminal button's cell membrane are pinched off and then recycled to form new vesicles by *cisternae* (a part of the Golgi apparatus) (see figure 8e).

figure 8e

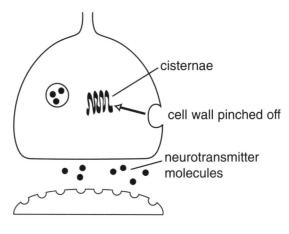

15
Outcomes: The Fifth Step in Neurotransmission

Once neurotransmitters have been released into the synapse, several outcomes are possible.
a) Neurotransmitter molecules can *bind* (combine) with post-synaptic receptor molecules to excite or inhibit the receiving neuron's firing action (see figure 9a).

b) Neurotransmitter molecules can be taken back up into the sending neuron, a process called *reuptake* (see figure 9b). Reuptake occurs almost continuously and accounts for most of the removal of neurotransmitters from the synapse.

c) Some of the reuptaken neurotransmitters *move back into vesicles* and are recycled through the process again (see figure 9c).

d) Other reuptaken molecules don't make it into the vesicles because they *get broken down by the action of an enzyme* (let's call this *enzyme 3*) into end products of chemical degradation, called metabolites. Now they're out of circulation. They're not neurotransmitters anymore (see figure 9d). The metabolites are

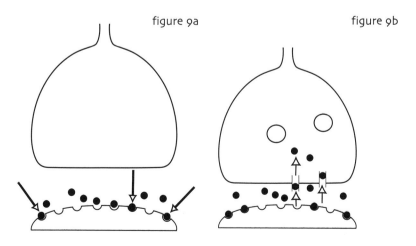

figure 9a   figure 9b

figure 9c

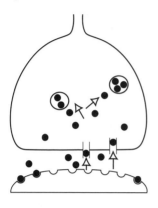

removed through the blood stream or reabsorbed in some other way.

e) Still other neurotransmitters don't make it out of the synapse at all. They *get broken down by the action of an enzyme in the synapse* (let's call this *enzyme 4*) into end products called metabolites. Now they're out of circulation too. Putting neurotransmitters out of circulation by breaking them down into metabolites is called *inactivation* (see figures 9d and 9e).

figure 9d

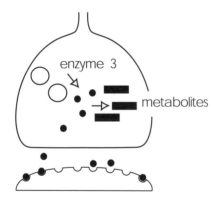

Although the sequence of events just outlined is the typical route for many neurotransmitters, there are several other possible routes. For example, serotonin and norepinephrine are released through varicosities (swellings) in branches of the axon. Some of them are released near postsynaptic membranes, so they appear to behave as in the process just described. Others are released diffusely throughout the extracellular region, thus acting as neuromodulators.

figure 9e

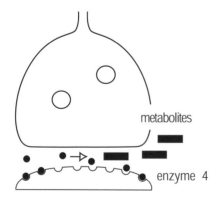

Peptides make a different voyage. They are manufactured in the cell body rather than in the terminal button, and are transported, inside vesicles, down the axon to terminal buttons. The biological activity of each peptide depends on its particular sequence of amino acids as well as on which of three kinds of receptors it attaches to. After release from terminal buttons, neuropeptides are not reuptaken and recycled. Those that don't reach their receptors are destroyed by more enzymes.

16
Some Classes of Drugs and Other Substances: How They Act in the Brain

Psychoactive drugs (legal and illegal, therapeutic and recreational) and other chemical compounds can interfere with normal processes of neurotransmission at any stage in the life span of neurotransmitters. The actions of specific drugs will be discussed in more depth in Part IV (Major Psychiatric Disorders) and Part V (Substance Abuse and Addiction). Here we introduce some important actions of different classes of drugs.

Drugs can:

a) *interfere with the synthesis of neurotransmitters.*
Example: low levels of lead, ingested over time, impede the action of enzymes that promote the synthesis of neurotransmitters from precursor substances (see figure 10a).

b) *facilitate or inhibit the release of transmitters.*
Example of a release facilitator: amphetamine promotes release of dopamine into the synapse. Example of a release inhibitor: calcium channel blockers, such as verapamil, sometimes used

Table 3

Categories of Therapeutic Drugs	Conditions for Which Drug is Used
Neuroleptics (antipsychotics)	schizophrenia, and as adjuncts in other conditions
Antidepressants	depression, obsessive-compulsive and eating disorders, aggressivity, attention-deficit hyperactivity disorder (if stimulants can't be used), some addiction cravings
Mood stabilizers (lithium) Anticonvulsants Calcium channel blockers	bipolar and schizoaffective disorders, episodic aggression
Anxiolytics (anti-anxiety)	anxiety
Psychostimulants	attention-deficit hyperactivity disorder

to treat bipolar disorder, affect the action of the calcium by blocking the channels on the cell membrane that allow calcium ions to pass through (see figure 10b). If the calcium ions can't get into the presynaptic neuron, vesicles won't fuse with the presynaptic membrane and won't be able to dispatch their loads of neurotransmitters into the synapse.

c) *potentiate (increase, strengthen) the action of transmitters.* Example: opiates such as heroin and morphine have molecular structures similar enough to structures of natural neurotransmitters (endorphins, enkephalins) that they can fool natural opioid receptor molecules into thinking they are the natural transmitters. "Opioid" refers to natural transmitters in the brain, whereas "opiate" refers to drugs. Opiate drugs work by combining with the same receptors used by the natural opioids. These drugs behave like natural opioid transmitters but in stronger concentrations than the natural chemicals, to give a heroin rush or to numb pain from an injury or illness. Opiate drugs act in similar fashion to the natural opioid transmitters, thereby adding to the action of the transmitter. In this situation, you have natural transmitters doing their stuff plus opiate drugs doing the same stuff. Drugs that increase the effects of another substance are called agonists (see figure 10c).

d) *block action of transmitters at postsynaptic receptor sites.* Example: *neuroleptics,* also called

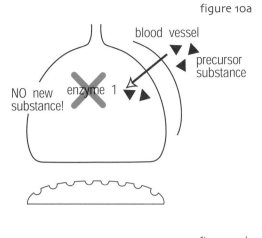

figure 10a

blood vessel

precursor substance

NO new substance! enzyme 1

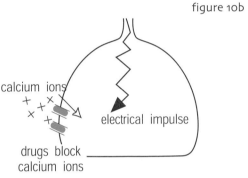

figure 10b

calcium ions

electrical impulse

drugs block calcium ions

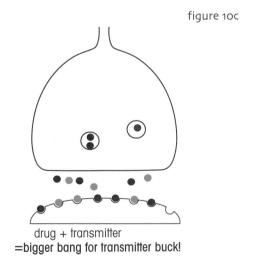

figure 10c

drug + transmitter
=bigger bang for transmitter buck!

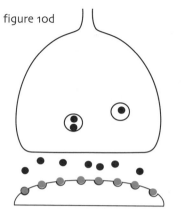

figure 10d

drug won't let transmitters into their postsynaptic homes

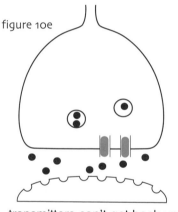

figure 10e

transmitters can't get back up!

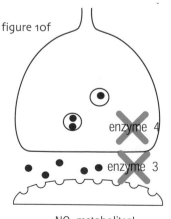

figure 10f

NO metabolites!

antipsychotics, include newer drugs that block both serotonin and dopamine (called *serotonin-dopamine antagonists,* or *SDAs*), and older drugs that block dopamine only (called *dopamine antagonists*). The latter group, including chlorpromazine (Thorazine), haloperidal (Haldol), fluphenazine (Prolixin), and many others, were the primary tools for reducing symptoms of schizophrenia until the 1990s.

The newer serotonin-dopamine antagonists, also called *atypical antipsychotics*, include risperidone (Risperdal), clozapine (Clozaril), olanzapine (Zyprexa), and others. They are now used more often than dopamine antagonists because they not only can ameliorate a larger range of symptoms, but also have a much more favorable side-effect profile.

These drugs also fool receptors. However, unlike the previous group of drugs, they do not act like the natural transmitters they resemble. They grab receptor slots away from the natural transmitters by binding more quickly and more tenaciously to the recep-

96

tors, thereby preventing the natural transmitters from doing their work. This postsynaptic process is called "competitive binding at the receptor site." Drugs that diminish the effects of other substances are called *antagonists* (see figure 10d).

e) *block reuptake of transmitters into the presynaptic neuron.* Example: selective serotonin reuptake inhibitors (SSRIs) block the reuptake of the neurotransmitter serotonin into the sending neuron, thereby increasing the amount of serotonin available in the synapse to combine with receptors on postsynaptic membranes. The SSRIs relieve symptoms of depression, obsessive-compulsive disorder, and other psychiatric conditions by remedying the inadequate supply of serotonin that is causing the symptoms (see figure 10e). These drugs have come into widespread use since the beginning of the 1990s, replacing older drugs (such as *tricyclic antidepressants*) that had multiple unpleasant side effects. Well-known SSRIs are fluoxetine (Prozac), sertraline (Zoloft), and paroxetine (Paxil). Sometimes these drugs have the specific troublesome side effects of preventing orgasm or decreasing libido, so it may be necessary to substitute other drugs.

Some drugs also alleviate depression by blocking reuptake of norepinephrine as well as serotonin into the presynaptic terminal, such as venlafaxine (Effexor), mirtazapine (Remeron), bupropion (Wellbutrin), and nefazodone (Serzone). Cocaine and its relatives achieve their effects by blocking reuptake of dopamine. Reuptake blocking of these different transmitters increases the amount of neurotransmitter left in the synapse to combine with postsynaptic receptors to give symptom reduction or, in the case of recreational drugs, a high.

f) *prevent inactivation of transmitters by enzymes.* Some enzymes promote the degradation (breaking down) of neurotransmitters into the end products called metabolites. When levels of a transmitter are too low, inhibiting the enzyme that is breaking the transmitter down increases the supply of that transmitter in the synapse. As a result, more is available to combine with postsynaptic receptor molecules.

Example: monoamine oxidase (note ending -*ase*, designating an enzyme) breaks down (inactivates) monoamine neurotransmitters. Monoamine oxidase inhibitors (MAOIs) block the action of the enzyme monoamine oxidase. They stop it from breaking down monoamine transmitters. Thus they increase the amount of monoamines available to do work (see figure 10f). Serotonin, dopamine, and norepinephrine are major monoamine neurotransmitters (see Table 2, p 77). The MAOIs relieve symptoms of depression related to inadequate supply of serotonin and norepinephrine by blocking the breakdown of these transmitters.

These monoamine neurotransmitters have a great deal to do with psychic activity. They are extremely versatile and play multiple roles in our lives. For example, dopamine is active in the pleasures of eating, drinking, and sex. It combines with several different dopamine receptors. For details about the talents of these transmitters, see Chapter 12.

Do you enjoy these? Thank your dopamine!

17
Using Neuroscience Information to Empower
Excerpt from Process Recording

Jose, age 38:
Angry, Abused, and Court-Mandated
Joseph D'Amico, MSW

Brief Profile. Jose is a thirty-eight-year-old Caucasian male who was mandated to attend individual psychotherapy by the court. Jose had gotten into an altercation with his girlfriend's ex-husband and was subsequently arrested for breach of peace. He is very loyal to those he considers his friends and his self-image is built upon his word. If he makes someone a promise, he will keep it—no matter what the cost.

Brief History. Jose is one of two children born into a family with a history of alcoholism and depression. His earliest memories, of which he can barely speak, involve being sent to foster homes, juvenile facilities, and physical and emotional abuse by his father. Between foster homes and juvenile detention centers he "ran the streets" where he learned how to fight and survive. He left school before graduating, drank, did drugs, worked at various jobs, and became a semi-pro boxer. Jose is from New York and came to his current residence in New England approximately seven years ago.

Treatment Background. The first several sessions with Jose consisted mainly of getting to know him and, thus, building a therapeutic relationship. Engagement was easy as Jose is a bright and very likeable person. Although his attitude appeared to be totally negative in that he was quick to anger and his speech was interspersed with swearing, it was not difficult to ascertain that this was his way of socializing. Jose had a volatile temper and would go from making a joke to immediately switching to anger. Once he was "set off" he would begin to rant and rave in a loud voice, loudly swearing and gesticulating with his arms and hands.

Session #4
Jose enters my office and sits down and we exchange greetings.

T: How are you doing, Jose?

F: Lousy. My boss is screwing me around and has me driving all over the place and owes me back money.

T: Oh. [A pause ensues with room for Jose to continue.]

F: You haven't heard the worst of it! I went to make a delivery in Bridgeport and pulled in where it said trucks enter. I blew my horn and waited. Nothing happened so I kept blowing my horn and finally this big fat B.... comes out and starts yelling that deliveries are around the corner. So I drive around the corner and find the entrance. I walk in and there's a guy standing behind a cage and I ask if this is where I'm supposed to drop off my supply. He gave me some smart answer and I lost it. I told him I'd jump over that cage and rip his arms off as well as that rotten tie he had on. [I'm omitting all the swearing that went with this.]

T: [Jose went on with this for a few more minutes before I intervened.] So, Jose, who is it that's so angry?

F: What do you mean?
T: I mean that I'm listening to a guy called 'Mr. Anger,' and that's not you. It seems to me that when Mr. Anger comes, Jose leaves. Make any sense to you? [Jose takes a few moments before answering.]

F: Yeah, it sort of does. When I get angry I seem to disconnect from myself, then nothing else matters. I'm so worked up that not only does nothing else matter, but all that exists is pure, blind rage.

T: We don't have to name Mr. Anger as he isn't any one person or event. Rather, he's a totality of years of trauma or abuse in various forms as well as ingested parts of family, friends, and environment. Now I realize that you truly want to get a handle on your temper as you're tired of losing it and then going through the emotional torment of feeling guilty and ashamed of your behavior. What I'd like to do now is explain what it is I think is going on in your brain. Sometimes just knowing that what is happening is grounded in your body and your brain and can be explained logically and scientifically makes a difference.

F: That would be great!

T: Well, to begin with, Jose, you have to sort of stretch your imagination. Our brains have been evolving for thousands of years and are pretty complex. The part of the brain that we're concerned with here is a part called the limbic system. This is an inner section of the brain, and animals that predated humans had limbic systems. What I'm alluding to is that this is a much less developed part of the brain but it is very important because it is the part that regulates things like hunger, thirst, sex, as well as emotions and passions such as love, joy, fear, sadness, and rage. This limbic system also has other functions including transmitting messages of pleasure and pain into memory. These memories are put into storage by a biochemical process that results in actual changes in microscopic brain structure. Whew…break time.

F: Joe, I'm not sure I got all of that…

T: I'll bet as I'm giving you a lot of information pretty quickly. Let's look at it from another angle. Being that our main concern is your anger, let's just focus on one part of the limbic system called the amygdala. The amygdala mediates rage reactions. When you get angry the amygdala is really revved up. If we take this one step further, we know that you have memories in storage in your brain that get activated by cues that at times you may not be consciously aware of. Since you have a trauma history, Jose, you have loads of repressed memories—that is, memories of things you aren't even aware of—and your buttons can get pushed a hundred times a day. The amygdala is going into overdrive. Make any sense?

F: Yeah! I get the gist of it, I do know what you mean. But now, how do I deal with it?

T: Well, perhaps just having some information is the first step. You've got plenty to reflect on till our next session.

References Part II

Carlson NR (2001). Physiology of Behavior, 7th ed. Needham Heights, MA: Allyn and Bacon.

Clark DL and Boutros NN (1999). The Brain and Behavior: An Introduction to Behavioral Neuroanatomy. Oxford: Blackwell Science.

Edelman GE (2004). Wider than the Sky. New Haven: Yale University Press.

Hyman SE (1997). Commentary. Science and Treatment (video). Produced by the National Alliance for the Mentally Ill, Arlington, VA, in conjunction with the National Institute of Mental Health, National Institutes of Health, Rockville, MD.

Purves D, Augustine GJ, Fitzpatric D, Katz LC, LaMantia AS, McNamara JO, and Williams SM (2001). Neuroscience, 2nd ed. Sunderland, MA: Sinauer Associates.

Purves WK, Sadava D, Orians GH, and Heller HC (2001). Life: The Science of Biology, 6th ed. Sunderland, MA: Sinauer Associates.

Sadock BJ and Sadock VA (2003). Synopsis of Psychiatry, 9th ed. Philadelphia: Lippincott, Williams, and Wilkins.

Winslow JT, Noble PL, Lyons CK, Sterk SM, and Insel TR (2003). Rearing effects on cerebrospinal fluid oxytocin concentration and social buffering in rhesus monkeys. Neuropsychopharmacology.28(5):910-8. Epub 2003 Mar 26.

Young LJ, Lim MM, Gingrich B, Insel TR (2001). Cellular mechanisms of social attachment. Hormonal Behavior 40(2):133-8. Review.

PART III
CHILD AND ADULT DEVELOPMENT

PART III
CHILD AND ADULT DEVELOPMENT

Advances of the past two decades have enabled new opportunities to integrate biological measures into developmental research. . . . . At the cutting edge of this paradigm shift are behaviorally oriented scientists testing biosocial alternatives to traditional models of development.
        Douglas A Granger and Katie T Kivlighan, 2003.

The scope of human development research has expanded greatly in response to the emergence of neuroscience. Biopsychosocial interactional models, representing changing configurations of individual-environment exchanges through time, have become the gold standard for child development research. In Part III, I review research on some of the major topics pertinent to human development. I have chosen a few topics and have considered a few issues that appear critical for human service practitioners. Many important issues have been omitted. The field of human development is so broad and so extensively researched that even lifelong specialists in the area would be hard put to carry out a comprehensive and exhaustive review. These brief chapters should be understood as simply an introduction to a few highlights.

Jerome Kagan, arguably the preeminent developmental psychologist of the second half of the 20th century, has summarized some of the conceptual changes in developmental research during the period from the 1950s to the early years of the 21st century. What was the situation in developmental psychology in the mid-20th century? In Kagan's words, "I blush as I recall telling my first undergraduate class in 1954 that a rejecting mother could create an autistic child" (2003, p. 3).

Two schools of psychological thought dominated American psychology during the decades of the mid-20th century: psychoanalytic and ego psychological theory, on the one hand, created and elaborated by Sigmund and Anna Freud and their followers, and theories of respondent and operant conditioning in the tradition of Watson, Pavlov, and Skinner, on the other. Proponents of ego psychology and early behaviorism vehemently denounced each other. Despite the marked differences between them, however, they had one thing in common: adherence to relatively simple basic assumptions and organizing constructs. The psychology of the day was seldom plagued with a sense of self-doubt. There were truths about human psychology, and you could access them (whichever truths you preferred) by mastering basic principles.

During the final three decades of the 20th century, systems theory asked the two schools of thought to account for influences from the world beyond family, friends, and colleagues. Economic forces, political power, cultural diversity, and organizational structures were recognized as critical influences on human behavior and development. Cognitive psychology challenged behaviorists to expand their framework to incorporate thinking and emotions. Social learning theory gained equal status with concepts of respondent and operant conditioning as an explanation for learned behavior. Finally, neuroscience emerged as a major contender in the last decade of the 20th century.

There was a disturbing consequence of this proliferation of theories and research: the comfortable days of certainty and confidence were gone, possibly forever.

Some important conceptual advances accompanied the increasing attention to cognition, emotions, environmental forces, and neurobiology. The role of temperament in development began to be recognized. Interdisciplinary research uncovered important information about critical periods in development and the role of brain maturation in emotions and cognition. Our understanding of psychological phenomena such as attachment, bonding, and affiliation advanced with new discoveries about their neurobiological foundations, enabling investigators to "peer beyond the behavioral display to infer more fundamental processes, as the student of musical composition perceives the theme hidden in the surface improvisation" (Kagan, 2003, p 3). Knowledge increased about the effects of stress, vulnerabilities, and resilience on children's development, mediated by invisible but powerful events in the brain. Finally, research on the neurobiology of consciousness produced intriguing insights into mind/body connections.

18
Temperament, Genes, and Development

What do genes have to do with social behaviors? A relatively simple example from animal research offers one answer to this question. Prairie voles (small mouselike rodents) make a monogamous attachment to a mate, but montane voles do not. Insel and Hulihan (1995) discovered that this difference in behavior between two otherwise similar animals is due at least in part to a small segment of DNA (deoxyribonucleic acid, a genetic material) in a gene responsible for the distribution of receptors for the neurotransmitter vasopressin in male prairie voles. Here, specific *brain structures* (a segment of DNA and vasopressin receptors) affect *physiological process*, by influencing activity of the neurotransmitter vasopressin *to produce the behavior* of having sex with only one mate! In female prairie voles, oxytocin rather than vasopressin is the monogamy neurotransmitter. Oxytocin is released by the female when she mates, an event that precedes monogamous pair-bonding.

What role did the voles' environments play in the emergence of the two different species-typical behaviors? Montane voles who lack this fragment of DNA have multiple partners if the environmental condition permits it-that is, if other willing and eager montane voles are available. (But if no other mates were available, wouldn't the montane vole settle for one partner rather than none?)

Why should we care about this piece of vole lore? Does this tidbit from animal research mean that some people but not others have a gene that predisposes to monogamous behavior? We don't know! But the effects of socialization are presumably more powerful in humans than in voles. For example, environmental influences probably explain a larger percentage of the conditions of sexual behavior in humans than in voles (epigenetic factors are more influential than simple genetics). Be that as it may, in simpler species and complex primates alike, many behaviors, feelings, and cognitive styles are strongly influenced by specific genes, with the particular *forms* these behaviors or feelings take being shaped by the environment.

A dimension of human behavior and development heavily influenced by genes has been called *temperament.* Temperament refers to a stable, biologically based profile of *mood* and *behavior* that emerges early in development. Some contemporary developmental psychologists restrict the definition to behavioral components only. For example, Zeanah and Fox call temperament a "behavioral style exhibited by infants or young children in response to a range of stimuli and contexts (2004, p. 33)." In our discussion, we will assume that temperament comprises both components (mood and behavior), and temperament will be considered as a lifelong aspect of individuals rather than primarily a characteristic of infants and young children. The concept of temperament has been valuable as it offers a possible explanation for the consistency of certain behaviors through time.

Children's temperaments vary greatly. (How many times have you heard mothers with more than one child say "They were as different as night and day from the minute they were born!") As we shall see, these individual differences play a critical role in development.

On the basis of their famous longitudinal study, Thomas and Chess (1977) identified nine temperamental characteristics: activity level, eating and sleeping rhythms, tendencies to approach or withdraw in novel situations, adaptability to changes in routine, emotional intensity, responsiveness to stimuli, mood, persistence, and distractibility or soothability. They found that the temperamental characteristics of individual children showed considerable stability into adulthood over a 25-year period. The advances in knowledge about temperament made possible by neuroscience findings have led various scholars to revise this formulation of temperament, although the dimensions identified by Thomas and Chess overlap with many of the more recent definitions.

A central concept of many of the newer definitions is that any temperamental category implies *specific anatomical structures* and *specific physiological processes* coupled with a *set of possible behaviors.* Behaviors chosen from this set depend on inputs from the environment. That is, temperament confronts the environment and influences the individual's choices about how to respond to the environment.

Mary K Rothbart, a leading temperament scholar, defined temperament as constitutionally based differences in *reactivity* and *self-regulation* (Rothbart, 1989). Reactivity is the ease with which responses are aroused in four areas: motor (by action of muscles), affective (emotional), autonomic (pertaining to involuntary, automatic body functions), and endocrine (through hormone-secreting glands). When we speak of self-regulation, we mean behaviors that *modulate reactivity,* such as approach, withdrawal, attack, inhibition, and self-soothing. As we saw in Chapter 7, genetically based characteristics shape the person by interacting with environmental inputs through complex adaptive systems. In Rothbart's model of temperament, individual differences in reactivity are present at birth—that is, they are strongly influenced by genetics. Self-regulation emerges gradually from interaction between genetically based characteristics and the environment.

An example of the interaction of temperament with environment came from the work of van den Boom (1994). She assessed infants' irritability at 15 days of age and then measured the babies' attachment to their mothers at 1 year of age using

Ainsworth's Strange Situation Procedure (Ainsworth, Blehar, Waters, and Wall, 1978). Mothers were primarily of low socioeconomic status. Temperamentally irritable infants were far more likely to be classified as insecurely attached 1-year-olds; nonirritable infants, more likely to be classified securely attached at 1 year. However, parent sensitivity was not related to attachment status as measured by the Strange Situation. This finding suggested that genetic influences were more important determinants of attachment status at age 1 than the quality of parental responsiveness.

In a second study, van den Boom repeated the measures, but only with temperamentally irritable infants. The infant-mother pairs were assigned to a treatment group and a control group. Control-group subjects received no intervention and were simply reassessed at 1 year. The mothers in the treatment group were visited at home and were given instruction about how to soothe their babies and how to play with them.

At 1 year follow-up, infants in the control group were much more likely to be classified insecurely attached, whereas infants in the treatment group were just as likely to be classified securely attached as the *nonirritable* infants in the first study. It appeared that infant temperamental irritability may affect the quality of mother-infant interactions. Irritable, hard-to-soothe infants behave in ways that push their mothers away. Thus irritability may impede attachment by interacting with mothers' not knowing how to soothe their babies and/or the exhaustion and frustration involved in trying to care for an irritable baby. It appears that a negative feedback loop takes shape: irritable infants discourage mothers' attempts to cuddle, frustrated mothers withdraw, and the infants lose opportunities for attachment-promoting exchanges with the mother.

In a study of child fear and compliance, temperament/environment interactions were also important (Kochanska, 1993). Because individual differences in fear reactivity reflect the child's sensitivity to conditioned signals of punishment, very fearful children should be more sensitive to maternal demands for compliance than less fearful children. Indeed, mothers'

112

directives elicited higher compliance in fearful children than in less fearful children. The child's temperament interacted with the mothers' management behaviors.

Behaviors related to cross-cultural differences in values and child-rearing attitudes were studied by Chen, Hastings, and colleagues (1998) using the Child-Rearing Practices Report scales. North Americans typically devalue temperamental *behavioral inhibition to novel stimuli* (limiting one's overt responses to the stimuli), but this is a valued characteristic in Chinese culture. Chinese children showed more behavioral inhibition to novel stimuli than their Canadian counterparts. In almost all cases, the direction of the relationship between the children's behavioral inhibition and the mothers' child-rearing beliefs was reversed between the two groups. In the Chinese sample, children's behavioral inhibition correlated positively with mothers' acceptance and negatively with a rejection and punishment orientation. In the Canadian sample, children's behavioral inhibition correlated negatively with mothers' acceptance and positively with a punishment orientation. The study showed the effects of prevailing cultural beliefs and norms on interactions between mothers' beliefs and children's behavior.

Kagan (1994) identified two clusters of temperamental qualities that usually persist from infancy through adolescence. Inhibited children (about 10% of American 2-year-olds) are described as shy, cautious, timid, watchful, and restrained. At the opposite end, uninhibited children appear naturally sociable, bold, energetic, spontaneous, and friendly. The majority of children fall somewhere between these two extremes.

Kagan and his colleagues observed two temperamental dimensions in the 2nd year as regards response to unfamiliar events: excessive shyness versus sociability, and timidity versus boldness. The researchers found that these characteristics in toddlers were related to characteristics that the children exhibited as 4-month-old infants, whom the researchers dubbed "high reactive" because the babies showed vigorous motor activity and distress in response to unfamiliar sights, sounds, and smells. In the 2nd year, the high-reactive infants tended to become shy, fearful, and timid in response to unfamiliar events.

The other group of 4-month-old infants ("low reactive") showed low levels of motor activity and little irritability in response to the same stimuli. These children were likely to become sociable and relatively fearless as toddlers.

The high-reactive children showed elevated levels of morning cortisol (a hormone produced by the body in response to challenge or stress), faster and less variable heart rates, and a pattern of right frontal activation on encephalograms (Kagan, Snidman, and Arcus, 1998). Kagan suspected that each temperamental type has a distinct neurochemistry that affects the excitability of the amygdala. Neurochemical profiles were thought to involve variation in the concentration of dopamine, norepinephrine, opioids, or GABA (see Table 2, p 77), or the distribution of receptors for these neurotransmitters. Kagan's hypothesis was that because GABA inhibits neural activity, infants who cannot regulate distress may have inadequate GABA function.

Some of the children who had been classified as high or low reactive at 4 months were studied again at the age of 11 years using a 3-hour battery of tests (McManis, Kagan, Snidman, and Woodward, 2002). A larger proportion of high- than of low-reactive children remained shy and subdued at age 11 years when presented with an unfamiliar stimulus, whereas a larger proportion of low than of high reactive 11-year-olds were sociable and emotionally spontaneous in the same unfamiliar situations. The 11-year-olds classified as high-reactive infants showed greater electroencephalogram activation in the right parietal lobe than in the left at age 11 and, if they had also been classified as fearful in the 2nd year, greater activation in the right frontal area (Fox, 1991). Together with other neurobiological measures, these observations suggest that greater psychological reactivity is associated with greater neurophysiological activity in very specific brain regions, and that observed behavioral differences first observed and recorded at 4 months of age were still present (although not universally so) when the children were 11 years old (Kagan, 2003).

A powerful implication of this longitudinal evidence for temperament is that children's temperaments can actually shape their environments. Since temperament is typically stable over the developing years, the behaviors that express the particular temperament are repeated over and over, and elicit similar patterns of responses over and over. A child's development is influenced by millions of exchanges back and forth between the child and his or her environment. Cheerful, relaxed, or outgoing infants and children tend to elicit from their environments responses that differ from those elicited by tense, anxious, or irritable infants and children. The developing child is affected by the feedback from its environment and then responds to it, so that there is an ongoing pattern of reciprocal influence. It's important to remember that a child's development is shaped not by temperament alone and not by environment alone. It's the repeated interactions between children and their environments—environments in which primary caretakers are very important but by no means the only actors—that are recorded in children's experiential memories and that form the basis for emotional learning.

Some aspects of the developing child's constitutional make-up can change at the neurobiological level as a result of the environment's responses to the child's responses. Kagan found that the temperaments of children born fearful and inhibited usually remained stable throughout development, but that in some cases, relaxed, low-key environments with specially sensitive mothers could lessen this inborn characteristic. In other words, environment can modify genetic effects. (A heavy trip for mothers? Yes indeed).

But these findings should not lead us to conclude that most children can be remade by parenting. Although temperamental characteristics can sometimes be modified, they tend to be stable. Parents influence, but they can't determine. With more knowledge and better support, parents can enhance and enrich, but they can't remake. American culture is permeated with the belief that parents have the power to raise their children to become happy, well-adjusted adults. The corollary of this belief is that adults who are unhappy must be the victims of parental failures somewhere along the way. Despite the fact that neither the belief nor its corollary is borne out by research evidence, the belief's continuing popularity, according to Kagan (1994), reflects the fondness of Americans for the idea of connectedness—that the past leads in lock-step fashion to the present.

19
Stress, Vulnerability, and Resilience

Stress. Several disciplines have contributed knowledge about the effects of stress on physical health, psychological well-being, and human development. People of all ages have varying vulnerability to stress and varying capacities to emerge from periods of stress relatively unscathed *(resilience).*

We know that egregious and long-term physical or emotional abuse in humans is associated with posttraumatic stress disorder, depression, and cognitive and emotional deficits. We now have a wealth of evidence about the neurobiological underpinnings of these effects. The connections between emotional neglect and these outcomes are less well documented in research literature, but few observers doubt that they exist.

Preliminary data suggest that severe emotional neglect (notably deprivation of caregiver/infant interactions over time) may cause abnormal metabolism in a part of the temporal lobe thought to be involved in social functioning. PET scans of eight children reared from infancy in Romanian orphanages until their adoption into American families showed these abnormalities. Although data to control for possible confounding variables were limited, such brain changes are believed to account for the difficulties some neglected children show in relating to people, despite extended periods in nurturing adoptive homes (Chugani, 1998). Some of these orphanages have been observed to provide almost no tactile or emotional stimulation between caregivers and children even when the facilities were found to be clean, well supplied with toys, and offering adequately nutritious meals (Talbot, 1998).

However, mild, moderate, or intermittent stressful events have not been demonstrated to lead to emotional disorders (except, perhaps, in uniquely sensitive individuals). By contrast, *chronic* or *extreme* trauma, stress, or deprivation are well-documented precursors to symptoms of psychiatric disorders.

Traumatic experiences have measurable effects on certain physiological indicators. Particular hormones, the glucocorticoids, are known to be central in the mediation of stress. Glucocorticoids produced by the adrenal glands (located on the kidneys), together with some of the monoamine neurotransmitters (dopamine, norepinephrine), comprise a frontline defense for mammals experiencing stressful conditions (Meaney, Diorio, Francis, et al, 1996). These substances mobilize the production and distribution of energy during stress through a feedback loop between the brain, the adrenal gland, and the external environment, known as the hypothalamic-pituitary-adrenal axis (HPA). HPA responses are genetically programmed but can be altered by early environmental events (Alves, Akbari, Anderson, et al, 1997). Prolonged bombardment of the human organism by these substances early in development can alter neurochemistry sufficiently to effect long-term changes in neurons and brain systems.

Glucocorticoids are so named because of their profound effects on glucose metabolism. Glucocorticoids break down protein and convert it to glucose, help to make fats available for energy, increase blood flow, and stimulate behavioral responsiveness. They decrease the sensitivity of the gonads (reproductive glands) to luteinizing hormone (LH), thereby suppressing the secretion of sex hormones.

Although glucocorticoids are essential for survival in the short-term, the adverse health effects of long-term stress are legion: high blood pressure, suppression of the immune system, damage to muscle tissue, infertility, inhibition of growth, slowed healing from injury, and brain damage (Carlson, 2001). Almost every cell in the body contains glucocorticoid receptors, so most of the body is affected by these hormones. When glucocorticoids are doing their stress-related work, there may be undesirable side effects. For example, male physician hospital residents showed severely depressed blood levels of testosterone, due most likely to their stressful work schedules, a response thought to be engineered by glucocorticoids (Singer and Zumoff, 1992).

The work of the HPA axis is probably the most researched example in the behavioral sciences of the complex adaptive networks that we encountered in Chapter 7. Let's walk through the steps that take place each time the system mediates stress-inducing influences from the external environment. These steps can help us understand the concept of complex adaptive systems by illustrating how one such system actually works (see Figures 11a and 11b).

1) An arousing event in the environment activates sensory cells in the eyes, ears, skin, nose, or mouth.

2) Sensory neurons in the peripheral nervous system transmit these records of a sensory stimulus from the receiving cells to the spinal or cranial nerves, and thence to the primary sensory cortex in the brain. Records of different kinds of sensory stimulation (such as warmth, pressure, or tickling, if the sense is touch), are then sent to different locations in the brain.

3) In the *sensory association cortex*, these records are converted into perceptions of sight, sound, touch, smell, and/or taste, then transmitted to the hypothalamus. The amygdala is also an important link between the cortical regions that process sensory information and the hypothalamus and brain-stem *effector* systems (systems that produce actions).

4) In response, the hypothalamus secretes a peptide called corticotropin-releasing factor (CRF), which is delivered to the nearby anterior pituitary gland by tiny local blood vessels.

5) CRF stimulates the anterior pituitary gland to produce adrenocorticotropic hormone (ACTH).

figure 11a   figure 11b

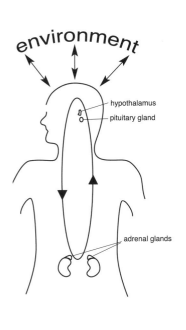

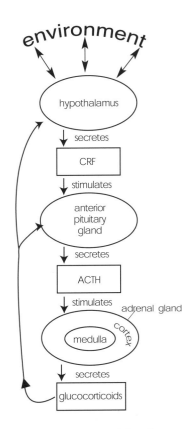

6) ACTH enters the general circulatory system (large blood vessels) and travels to the adrenal gland located on top of each kidney.

7) ACTH stimulates the adrenal cortex (outer layer of the adrenal gland) to produce the group of hormones called glucocorticoids.

8) Glucocorticoids are released from the adrenal glands and are dispatched around the body through the circulatory system.

9) These glucocorticoids produce the energizing effects described above, but also act to inhibit further production of glucocorticoids by negative feedback loops to the pituitary gland and the hypothalamus. That is, the HPA axis self-regulates through positive feedforward and negative feedback loops. When stimulated, the HPA axis releases hormones into

the bloodstream that have complementary but opposing functions (Granger and Kivlighan, 2003).

10) As the adrenal cortex is producing glucocorticoids, neurons in the adrenal medulla (inner section of the adrenal gland) are producing hormones that double as the neurotransmitters epinephrine and norepinephrine, found in the brain. Epinephrine (also called adrenaline) transmits messages to increase the heart rate and raise blood sugar. These actions of the adrenal gland illustrate the interweaving of brain functions with other body functions.

11) If stress-inducing events in the environment continue to bombard the individual, the cycle repeats, overriding the attempts of the negative feedback loop to reduce the secretion of more glucocorticoids. As these repetitions continue through time, at some unknown moment glucocorticoids around the body start to cause damage to body cells.

The importance of the actions of the HPA can hardly be overestimated. Hans Selye, a leading scholar of stress, suggested that most of the harmful effects of stress were the result of prolonged bombardment of the organism by glucocorticoids (Selye, 1976). His hypothesis later proved correct.

The early postnatal environment can contribute substantially to the development of stable differences in HPA responsiveness to stressful stimuli. Animals exposed to physical trauma, the administration of endotoxins (substances that make animals sick, such as salmonella), or maternal separation, showed increased HPA response to stress. These early effects can have deleterious effects later in life, especially in individuals genetically predisposed to stress-sensitive disorders such as anxiety, depression, and Tourette syndrome (Meaney, 2001; Alves, Akbari, Anderson, et al, 1997).

Evidence suggests that glucocorticoids acting through glucocorticoid receptors in the central nervous system may operate as lifelong organizing signals activated before birth and continuing to the end of life. When too many glucocorticoids bombard our neurons in old age, nerve cell death may be hastened

(Fuxe, Diaz, Cintra, et al, 1996). Overall, glucocorticoid receptors participate in neuronal plasticity (the capacity of neurons for change) from fetal and postnatal life to the onset of adult life and aging (Cintra, Bhatnagar, Chadi, et al, 1994).

Stress can also change developmental programming prenatally. Glucocorticoid receptors affect the regulation of the HPA axis during fetal development and can produce long-lasting changes in dopamine communication in the striatum and changes in serotonin communication in the brainstem (Slotkin, Barnes, McCook, and Seidler, 1996; Fuxe, Diaz, Cintra, et al, 1996). Stress-induced increases in maternal glucocorticoids may be a mechanism by which prenatal stress impairs the development of the adult offspring's glucocorticoid response (Barbazanges, Piazza, LeMoal, and Maccari 1996).

What environmental factors might diminish stress? Infant rats whose mothers engaged in frequent licking and grooming behaviors developed into adult rats showing smaller HPA responses to stress and decreased fearfulness in novel situations (Caldji, Diorio, and Meaney, 2000; Liu, Diorio, Tannenbaum, Caldji, et al, 1997).

Enriching rat environments also positively affected development. Rats housed in groups with ample opportunities to explore new objects and to interact socially increased their brain weight and cortical thickness and produced more new synapses, more complex dendritic branching, and more new blood vessels than did their less privileged peers (Black, 1998).

There's other good news. Environmental stimulation through postnatal handling or enriched environment can sometimes undo early damage. Specifically, in experiments with rats, impaired avoidance acquisition in adulthood caused by inadequate stimulation in early life was reversed by environmental enrichment (Escorihuela, Tobena, and Fernandez-Teruel, 1994). Either short periods of infantile handling or stimulation, or the administration of a drug that enhances $GABA_A$ receptor action, attenuated the adverse consequences of stress by decreasing HPA responsivity to stress (Meaney, Diorio, Francis, et al, 1996; Patchev, Montkowsi, Rouskova, et al, 1997).

As stress and traumas pummel developing individuals and

121

increase the risk of physical and mental disorders, biological and environmental protective factors fight back. The contest between risk factors and protective factors continues from conception to the end of life. (Did you ever think of your body as the arena for a mighty, never-ending battle?)

A major stressor for most young animals is the loss of, or separation from, caretakers, in particular a caretaker to whom the animal has formed an attachment. Human children and other young alike react with sadness, anger, apathy, and/or withdrawal to a parent's death, absence through separation, or emotional unavailability. Rhesus monkeys, deprived of contact with caretakers in infancy, show seriously deviant social behavior in later life (Harlow, 1958; Suomi and Harlow, 1972). So do some dogs (beagles) (Fuller and Clark, 1968).

The transmission of atypical neurochemistry from stressed mothers to their fetuses obviously involves a biochemical process. On the other hand, death or the prolonged absence of a parent is clearly an environmental condition. In both the biological and the environmental conditions, the involvement of the child's HPA axis over an extended time is a hazard to development. In the case of parental emotional unavailability, not uncommon when the caretaker suffers a mental illness or severe addiction, environmental stress on the child is cycled through the HPA axis and can subject the child to the long-term adverse effects of glucocorticoid overload, as reported above.

Maternal depression has been associated with changes in infants and children. Newborns whose mothers were depressed either before or after birth, or both, had elevated cortisol and norepinephrine levels, lower dopamine levels, and greater relative right frontal EEG asymmetry. Not surprisingly, effects on newborns appeared to arise more from prepartum than immediate postpartum maternal depression (Diego, Field, Hernandez-Reif et al, 2004). Ongoing maternal depression through the child's early years has been shown to exert small but significant effects on children's cognitive and emotional development (Beck, 1998).

Studies have shown that the role of maternal depression in the child's physiological and emotional distress is not unidirectional. It can act through different routes. (1) The mother suffers depression during the pregnancy, causing the child to be born with compromised HPA function (a physiological route of transmission). (2) The mother is depressed for reasons unrelated to characteristics of her child, but because of depression, she fails to transmit enough positive sensory and affective energy to the baby (through talking, smiling, making eye contact, cuddling), a psychosocial contributor to the baby's distress. (3) The mother's depression is a reaction to aspects of the baby, such as irritability, illness, or prematurity (a psychosocial contributor to the mothers' depression). (4) Gene(s) predisposing the mother to depression have been transmitted from the mother to the child (a genetic explanation). (5) Any or all of the above effects may be taking place simultaneously as interactive psychosocial and neurobiological events.

One remedy for lack of sensory inputs to infants that sometimes occurs when a mother is depressed appears to be breastfeeding. Breastfeeding is a multisensory experience—all five senses are involved. Since the physical benefits of these sensory experiences can be offered even by emotionally unavailable mothers, breastfeeding might relieve some negative effects of stress-induced glucocorticoids on the infant. Jones, McFall, and Diego (2004) looked at the association between breastfeeding, physiological measures of infants, and the ways mothers and infants related to each other. In this study, 78 mothers and their infants participated; 31 of the mothers were depressed.

Positive effects of breastfeeding were confirmed by the study. Depressed mothers who had stable breastfeeding patterns were less likely to have highly reactive infants. Infants of depressed mothers who breastfed did not show the frontal asymmetry patterns reported in previous studies. Moreover, breastfeeding stability, even in depressed mothers, was related to more positive mother-infant interactions. These findings suggested that promoting and supporting breastfeeding could enhance a positive feedback loop between infants and depressed mothers. It should go without saying that treating mothers' depression should be a top priority goal, for their own well-being as well as for that of their children.

The sense of touch has been used explicitly as a therapeutic intervention with several different populations. Regular touching has shown beneficial effects on the physical well-being, development, and emotional states of infants as well as older persons. Infants born to HIV-positive mothers were given three 15-minute massages daily for 10 days. These babies showed superior performance on almost every Brazelton newborn cluster score and had a greater daily weight gain than control-group infants (Scafidi and Field, 1996). Touch appears to be an entry point into the HPA-axis feedforward and feedback loops. For additional information on the use of touch, see Chapter 31.

The question of separation has concerned many a parent entering or reentering the labor market, whether by necessity or by choice. In particular, parents have struggled with the decision to entrust a child to group care by strangers. Issues of concern to the parents are: Will my child be safe? Will these teachers find time for her with so many other children to attend to? Is this a high-quality daycare center? How will the quality of care compare with what I can give at home? How will she feel and behave in a large group of children? Is the group care environment kind and gentle enough for her? What bad reactions is she likely to have? How will spending long days in care affect my child?

A second set of questions lurks behind the first. Am I being selfish? Will my child feel rejected? Am I really rejecting her? Will my family disapprove? Will my child be maladjusted later? Am I really doing this just to get out of the house and be with people? Am I a bad parent for foisting my kid on babysitters? Am I abandoning my child?

Many parents have observed that being dropped off at daycare stresses their children, causing the parents in turn to feel stressed. A recent issue of *Child Development* reports several studies on the short- and long-term effects of center-based child care on children's development.

During the early years of life, children's needs for social skills shifts from very limited to extensive, as they are increasingly required to interact with adults and peers in sophisticated ways (Johnson, Christie, and Yawkey, 1999). In full-day center-based child care settings, children must learn to navigate a complex social and cognitive environment with many same-age peers. It appears that children respond physiologically to this challenging environment with increased cortisol production. Cortisol is a glucocorticoid manufactured by the adrenal cortex. Cortisol levels typically rise in stressful or challenging situations.

Each of the studies focuses on one or two aspects of the complex influences of the daycare experience on children's development. When all the studies are taken together, the many relevant variables that are reported show promise in advancing our knowledge about the various influences. To answer the question "is this good for my child?" meaningfully, however, we must look at the *interaction* between the many important variables that function in the family's complex adaptive systems (see Chapter 7).

One of the studies assessed children's stress levels by taking samples of saliva and measuring the hormone cortisol at 10:00 a.m. and 4:00 p.m. from children in the daycare center, from the same children on days they stayed home, and also from a comparison group of children being cared for at home (Watamura, Donzella, Alwin, and Gunnar, 2003). Among children cared for at home, 71% of infants and 64% of toddlers showed decreases in cortisol levels from a high in the morning to the low in the afternoon, following the natural cycle of cortisol production. In the fulltime daycare centers, however, 35% of infants and 71% of toddlers had rises in their cortisol levels in the afternoon rather than lowering. What might this disparity mean?

Several possible explanations have been suggested. First, children's stress responses to full-time center-based child care differ, and these differences are associated with emotional tendencies that may precede their entry into care (Crockenberg, 2003). That is, *temperament* appears to affect the level of stress a child experiences. Children described by teachers as more socially fearful showed larger cortisol increases (Watamura et al, 2003). Temperamentally fearful children apparently show greater stress responses to center-based care where considerably more social interaction takes place than in the secluded environment at home.

Second, the *quality* of center-based child care does affect stress levels of children as measured in their cortisol levels. Preschoolers produced larger rises in cortisol when the daycare was judged to be of lower quality. However, many other risk and protective factors interact to impact the child.

Third, differences in children's aggression, *independent of the quality of caregiving* both at home and in the center, are associated with the amount of time spent in care. The more hours spent in daycare, the higher the cortisol levels and the greater the aggression. Yet the effect was quite small statistically. It appears that only some children, not all, are negatively affected by longer hours in group care (National Institute of Child Health and Human Development, 2004).

Several research groups are currently studying connections between early experience in center-based daycare, HPA activity, and behavioral outcomes. The researchers are testing interventions intended to prevent maladjustment in children at high risk for emotional or behavioral disorders. A 10-year study (1997-2007) is assessing the effects of a prevention program for low-income urban preschoolers at high familial risk for conduct problems (Brotman, Gouley, Klein et al, 2003). The researchers are evaluating effects of the prevention effort on affective and behavioral regulation as well as on physiology.

Children and families in the experimental prevention group receive multiple interventions related to parenting groups, pre-school groups, parent-child interactions, and home visits. Children's reactions to a challenging "ecologically relevant" peer situation are assessed prior to and after the peer experience. The test situation requires the children to enter a classroom where all the other children are already playing with each other. The situation is deemed ecologically relevant because it is an instance of a real life social challenge.

Experimental (intervention) and control children complete the task in the same way, and their behavior is rated by trained observers who do not know whether the child is in the intervention condition or control condition. Preliminary results show that intervention children enter the peer situation and play more interactively than control children, both by more engaging social behaviors and by an absence of withdrawal behaviors.

The cortisol levels of intervention children *increased* prior to entry into the challenge situation, and then declined after the peer entry situation to home resting cortisol levels. Control children, however, did not show rising cortisol in the challenge situation. Does this mean that intervention children are *more* stressed than control children in this situation? Apparently so—but cortisol increases are involuntary responses to challenge situations that help people cope. These children did so—by engaging. It appears that their HPA axes gave them the impetus they needed to succeed. The control children appear not to have experienced as much arousal, but rather continued to avoid engagement. This finding illustrates that rises in cortisol are not inherently bad—they help people meet challenges, and become harmful only if the level of arousal continues for too long a period

In a program designed to address the needs of preschool-aged foster children, called Early Intervention Foster Care (EIFC), Fisher and colleagues have been investigating whether improved behavior is matched by changes in neural systems involved in stress regulation, specifically the HPA axis.

Using the foster care setting as the milieu for therapeutic intervention, program staff members actively engage foster parents as therapeutic agents. Parents are trained to use consistent, nonabusive discipline, high levels of positive reinforcement, and close monitoring and supervision of the child. In a pilot study and a more recent large-scale randomized clinical trial, results indicate that children in the intervention group show significantly less disruptive behavior and less dysregulation in HPA axis activity than do children in the control group (Fisher, personal communication; Fisher, Gunnar, Chamberlain, and Reid, 2000; Fisher and Chamberlain, 2000; Fisher, Ellis, and Chamberlain, 1999). These results provide optimistic, albeit preliminary, evidence that neural systems affected by early stress may be responsive to environmental changes in early childhood.

Preliminary findings from studies by Fisher and colleagues and Brotman and colleagues suggest that environmental factors, operating at key points in development, may shape affective and behavioral regulation in human children, with corresponding changes in HPA axis function—exactly the process of biological and psychosocial interactions framed by the complex adaptive systems model (see Chapter 7).

Several interesting questions arise. Do changes in patterns of behavior and cortisol levels that occur in specific contexts endure over time? What additional factors such as temperament, age, previous experience with group settings, and cognitive appraisals, affect outcomes? What physiological indicators beside cortisol levels might be needed to confirm the findings? What are the roles of individual differences when children in the same context respond differently?

Do the patterns shown by relations between behavior, cortisol levels and context confer risk or resilience? Are rises in cortisol in the group contexts harmful, or are they positive adaptive responses to the cognitive and social challenges of group care? Will the early challenge in social engagement simply make it easier for children to adjust later to school, with corresponding declines in cortisol levels?

What about the danger of elevated cortisol associated with mild repeated stress related to center-based daycare? Is there evidence of any harm to developing brains? Hopefully not, because the daily experience of producing mild elevations in cortisol at child care did not appear to produce a permanent change in daytime cortisol rhythm. The children reverted to normal rhythms on days when they were not in child care.

The most likely consequence is that daily increases in cortisol may contribute to the heightened susceptibility to illness that is well documented among toddlers in child care. Cortisol is known to dampen activity of the immune system, increasing the likelihood that exposure to viruses will produce illness. Negative impacts on social or cognitive development seem unlikely, given the overwhelming evidence from studies of center-based child care showing that these settings, when they are of good quality, stimulate cognitive and social development (Watamura et al, 2003).

However, the impacts of lower quality child care on development are unclear. There is some emerging evidence that long hours in child care may be associated with increased externalizing behavior, especially in the case of center-based child care (Belsky, Weinraub, Owen, & Kelly, 2001). Nonetheless, without knowing whether cortisol levels that rise over the child care day do confer long-term harm on developing neural systems, it's obviously wise to assess the factors that contribute to the child's stress and consider whether and how they can be modified. This inquiry may be particularly important for temperamentally fearful children who may be susceptible to larger increases in cortisol over the child care day.

Newcombe (2003) identifies limitations of the studies due to omission of potentially important variables. Variables of interest are sometimes artificially isolated from other variables that actually operate in the context, so that some of the situations studied may be ones that do not exist in the real world. For example, amount of time in child care as a contributor to cortisol elevations was studied by controlling for family income and maternal depression. However, use of child care is also closely linked with maternal employment, which *increases* family income and *decreases* maternal depression. These variables (family income and maternal depression) in turn are linked to children's socioemotional adjustment. Therefore failure to include these variables in the analysis may give inaccurate estimates of real life effects.

Newcombe's insights underscore the need not only for prospective longitudinal studies, but also for the incorporation of models that trace interactive effects of several variables through time. In addition, future studies will continue to require interdisciplinary input, to carry out comprehensive assessment of complex biopsychosocial phenomena by measuring physiological, neurobiological, behavioral, and environmental processes within a developmental systems model. Furthermore, experimental manipulations of the kind evaluated in the studies we've reviewed are necessary to evaluate causal relations between the environment, child HPA axis function, and later behavior. If research on preventive measures shows that these interventions can improve outcomes, as preliminary results indicate, then it may teach us how to interrupt negative feedback loops and initiate positive sequences in their place.

Resilience. Not all individuals are harmed by early deprivation. Rhesus monkeys and beagles are harmed, but crab-eater monkeys and terriers aren't (Fuller and Clark, 1968)—another example of evidence that interaction between an infant's biological qualities and its life experience determine outcomes. Genetic differences among species of monkeys and dogs apparently create temperamental characteristics that lead them to process the experience of caretaker deprivation in different ways. Humans exhibit these differences as well (Kagan, 1994).

Several studies have shown that psychological damage early in life is often reversible, and that the degree of long-term harm conferred on children by early deprivation may vary considerably with the child's temperament. Most notable are the longitudinal studies of Emmy Werner (1989), who followed more than 200 children on a Hawaiian island from birth through the age of 32. Some of the children reared in horrendous circumstances were doing well at age 32. Conversely, children who appeared to have had every material and emotional advantage during childhood sometimes had emotional disturbances and/or tumultuous lives at age 32.

A growing body of research focuses on resilience, people's ability to withstand or bounce back from adverse events. Taken as a whole, recent research on child development and adult psychopathology discredits long-popular developmental theories which presuppose that progression is continuous through various psychological stages. According to such theories, in order to progress satisfactorily to the next stage of development, a child must "pass" the current level—no skipping allowed (Kagan, 1994).

With respect to advancing through the presumed stages, these theories use phrases such as "successfully navigate" or "resolve conflicts inherent in" or "achieve the tasks of" a particular stage in order to make a successful transition to the next stage. Does failure to negotiate developmental tasks associated with an oral, anal, or genital stage by the developing child turn into a personality conflict or a personality deficit, often repressed, that resurfaces in adulthood as a psychological disorder? The evidence runs counter to this belief.

Numerous instances of resilience illustrate the self-righting tendencies of the human species. For example, children adopted into American families after early rearing in East European orphanages showed markedly diverse outcomes. At the time of adoption, children were often cognitively impaired, had visual problems, sensory integration deficits, and language-processing difficulties that impeded cognitive and emotional functioning. However, in one study of more than 200 children adopted into American families, as many as 60% made impressive leaps forward in cognitive and emotional development during the years following adoption. Another 20% continued to have serious deficits. The remaining 20% (dubbed "resilient rascals") showed no ill effects from their institutionalization (Groza, Ryan, and Cash, 2003; Talbot, 1998).

Observers agree that the child's age at the time of adoption is a critical factor in the likelihood of normal development. Yet even children adopted with major deficits and at older ages often show impressive developmental catch-ups (Rutter, 1998). Another study concluded that all 46 adoptees from a Romanian orphanage had been able to form some sort of attachment to their adoptive parents (Talbot, 1998).

A study of resilience in 205 children from an urban community sample assessed children in elementary school and then followed them for 10 years (Masten, Hubbard, Gest et al, 1999). Outcomes of competence in late adolescence were examined in relation to adversity over time, preexisting competence, and psychosocial resources. Better intellectual functioning and parenting resources were associated with good outcomes across competence domains, *even in the context of severe, chronic adversity.* IQ and parenting resources appeared to have a specific protective role with respect to antisocial behavior.

The resilient adolescents had adequate competence across the three domains of academics, conduct, and peer relations despite a history of high adversity. They had much more in common with their low-adversity competent peers, including average or better IQ, supportive parenting, and psychological well-being, than with peers having high-adversity histories similar to their own, who had few resources and high negative emotionality.

In a sample of middle-class and working-class families with normally developing children and adolescents, Booth, Johnson, Granger, and colleagues (2003) assessed the expression of testosterone-related problem behavior. In adolescents, high levels of testosterone are related to stage of pubertal development and have been linked to problems with aggressivity, risk-taking, and depression. The investigators found that increased problematic risk-taking behavior and increased levels of depression were related to lower quality of child-parent relationships. The higher the quality of child-parent relationships, the lower the amount of testosterone-related adjustment problems.

The authors present these findings as an example of the mediating influence of psychosocial factors on a known biology-behavior connection. However, alternative explanations fit equally well with an integrated biopsychosocial model: could more positive child-parent relationships be a *response to* more pleasant, unproblematic behaviors by the children rather than a *stimulus for* them? Or could child-parent interactions form two different kinds of feedback loops in which hormone-driven troublesome behavior was either amplified or reduced in response to parental behaviors?

Granger and Kivlighan (2003) note that biological functions set the stage for behavioral adaptation to environmental challenge, and that environmental challenge, in turn, may affect fast-acting biological processes (such as hormone secretion) and slow-acting biological processes (such as gene expression). Biological activities that facilitate a particular behavioral process may be either stimulated or attenuated by social forces. Once again, the theory of complex adaptive systems is invoked as it integrates inborn temperament, biological processes, and social and environmental forces in developmental psychology.

20
Critical Periods in Development

The concept of critical periods in development is not new. In the 1950s, the ethologist Konrad Lorenz delighted scholars of behavior with pictures of goslings toddling and swimming after him. They had adopted him as their mother. Newly hatched goslings experience a critical period for attachment during the first few hours after breaking out of their shells. They follow the first moving creature that they see and hear, a phenomenon called "imprinting." Having seen and heard Lorenz before any other creatures, the goslings preferred him to a goose mother! Once imprinting occurs, it is irreversible. However, if goslings are not exposed to an appropriate stimulus during this short time period, they never form typical parent-child relationships. During the first week of life, infant rats develop a lifelong preference associated with the odor of their mother's nipples. Oxytocin speeds up the conditioning process for maternal odor cues, and blocking oxytocin delays the conditioning process (Nelson and Pankepp, 1996).

For birds, early imprinting plays a role in social development and later sexual preferences. And the bonding process works in the opposite direction as well. Animal parents also form bonds with their young at critical, often very short, time periods. Ewes imprint on the scent of their own lamb for a period of 2-4 hours after giving birth. After this time, they rebuff approaches by other lambs.

Critical periods are windows of opportunity for learning skills. Once a given time period has passed, the nervous system often becomes refractory to further experience, and the process of learning a skill that formerly came naturally becomes slow and difficult. If we knew when and how these critical periods occur, we could target our efforts to take advantage of them.

The periods are limited to specific times and to particular species. They are not times of generally enhanced learning. Songbirds learn their songs mostly during the first 2 months after hatching. The juvenile male bird listens to the song of a nearby adult male, memorizes it, and then matches his own song to the memorized model through auditory feedback! Exposure to other songs after the sensory acquisition period does not affect this learning. The bird vocally mimics only those songs heard during the sensitive period. Birds retain this skill for months or longer even without further vocal practice. Juvenile males need to hear the male model's song only 10-20 times to be able to reproduce it months later. And when young birds are presented with tapes of songs of many bird species, juveniles mimic the songs of their own species. That is, the birds are innately predisposed to learn the songs of their species (Purves et al, 2001). Do humans also have innate predispositions to learn certain things? We know that's the case—some of us learn math easily; others of us pick up tennis in a couple of sessions on the court.

However, in human brain development, critical periods are the exception, not the rule. Thompson and Nelson (2001) prefer to call these "sensitive" rather than "critical" periods, because the periods for developing complex behaviors such as human language are much longer and less well delimited. The example of goslings, which have a critical period lasting only a few hours to form attachment to a parent, is not characteristic of humans. In humans, the process of child-parent attachment is far more complex and is not yet well understood. We do know that the process can extend for years.

We also know that brain development is nonlinear. That is, it doesn't progress in a straight, uphill direction. In most species, including humans, brain maturation involves phases during which there's a flurry of activity, followed by phases in which activity diminishes.

Brain development continues throughout life. Although it is certainly true, as the Birth-to-Three movement proclaimed, that the period from birth to age 3 is a critical time for brain development, so is the prenatal period, and so are the years from 4 to adolescence. In adulthood the process continues, albeit at a slower rate. We can still learn in our '80s and '90s, and each time we learn something, a miniscule change takes place in a structure in our brain.

A few years ago, media coverage of neuroscience discoveries about child brain development drew widespread attention and aroused consternation. Television gurus admonished parents to stimulate their children in every possible way to promote brain development. Otherwise, these alarmists cautioned, the children might be left behind on the long road to admission into Harvard or Princeton. Parents scrambled to talk more to their babies in the crib and in utero, and Zell Miller, then governor of Georgia, decided to send parents of every newborn in the state a classical music CD! The so-called Mozart effect had caught the imagination of the popular press.

Although sensory stimulation of young children is believed to be very important in promoting development, some scientists became concerned. Where was the attention to the potential and often actual damage to children's brains from prenatal malnutrition, poor health care, viral infections and other maternal illness, exposure to toxic substances through environmental contaminants, alcohol and other drugs, and chronic maternal stress? The child's brain in utero and during the early postnatal years has been shown to be extremely vulnerable to these hazards, yet the relevant knowledge from research, decades old, was being virtually ignored by policymakers and media gurus. Where was public concern about creating safe and healthy environments for all of America's children? So far, politicians have failed to answer these questions.

Contrary to prevailing belief at the end of the 20th century, and since the publication of the first edition of this book, neuroscientists have discovered that new neurons do grow postnatally. Neurogenesis (formation of new neurons) has been discovered in the hippocampus and possibly in the parietal and prefrontal cortices (Gould, Reeves, Graziano, and Gross, 1999). This revolutionary discovery offers new hope for recovery from brain injuries and illnesses. Although the new neurons have been found only in a few specific brain regions, more may be discovered. Even if not, the coming of age of stem cell research may at some time in the future give us more generic tools to create new neurons and brain structures lost through illness or injury.

However, it has long been known that neurons alone do not make things happen; connections are needed for various brain activities to occur. We've studied these connections through the functions of synapses (Chapters 11-16). At birth, there are relatively few connections between neurons compared with later in life. Over the first years of life, many connections are made, and then some are eliminated. Molecular changes take place when pathways are used that result in the hardwiring of some connections and the elimination of others. By adolescence, pathways that have been used repeatedly are hardwired, whereas processes that haven't reached a certain threshold of use are eliminated (use it or lose it).

Before birth, more neurons are produced than we need. After birth, connections proliferate richly for 8 or 9 years. This rapid formation of synaptic connections is followed by pruning of dendrites and elimination of many connections. The entire process of connection-building and pruning extends from infancy into adolescence. Even in the early teens, density of synapses can exceed that found in human adults.

Neurons in creatures raised in an enriched environment may be better protected from elimination, as well as forming more connections. For example, mice reared in an environment enriched with toys, running wheels, and other interesting things had a thicker cerebral cortex than mice reared in a more limited environment. The enriched mice showed not only more connections, but also more neurons (Volkmar and Greenough, 1972).

Patterns of development are a complex sequence of modulating events. Critical periods of change during the first 10 years of life give us opportunities to retain and increase the efficiency of connections through repeated use, and to eliminate connections that aren't used. This is nature's way of fine-tuning neuronal circuits. Ongoing and repeated interactions between the human organism and the environment sculpt unique individual neuronal architecture (Chugani, 1997).

Figure 12 shows levels of human neural activity in the cerebral cortex at different ages as measured by rates of glucose metabolism. Metabolic rates in various cortical regions can be identified using PET scan (see Chapter 4).

The different symbols on the graph indicate different regions of the human cortex (frontal, parietal, temporal, and occipital). The graph shows that the developmental pattern is similar for the four regions. Since 95 % of the glucose employed by the brain is used for connections, not for neuron cell bodies, PET scans can be used to estimate the number of synaptic connections present at different periods in development.

As the graph shows, the number of connections in the cerebral cortex increases dramatically in the first years after birth. Glucose values are about 30 % lower at birth than adult rates, and increase rapidly to reach adult values by the second year. By the third year, these values exceed adult values. By about 4 years of age, a plateau is reached that extends to 8-10 years of age. The peak level is more than twice adult levels. There appears to be a transient phase during child development when the brain requires more nutrients to support its activities than do adult brains. Thereafter, glucose values gradually decline and reach adult values by about 16-18 years. That is, changes in patterns of connections, as indicated by levels of glucose metabolism, are almost completed by mid-adolescence. The environment influences which connections are to be kept and which eliminated. Although we can't determine how neurons are shaped before birth, we can have some influence on the final connections that are made between neurons after birth.

Patterns of glucose utilization at birth differ markedly from adult patterns. In newborns, the most active regions are the sensory and motor cortices, the brain stem (region of the brain leading to the spinal cord), and parts of the thalamus and the cerebellum (see Figure 2, p 28). The cerebral cortex is not yet very active, although the cingulate cortex and the hippocampus are already busy regulating emotions. These patterns of glucose metabolism in newborns are in keeping with their relatively limited behavioral repertoires.

By 3 months, more regions are active. Several connections that mediate behavior have been added . For example, the infant can now track visually and can reach for and grab objects. At 7-9 months, the frontal cortex has become active. This development is associated with recognition of danger, often observed in babies as "stranger anxiety." In his cross-cultural studies, Kagan (1994) observed that in all cultures, infants develop stranger anxiety around the age of 8 months. Recent neurobiological research offers an explanation of why *8-month-old Susie, who used to love everyone, now screams at the nicest ladies!*

The frontal cortex is the latest structure to become active in the brain. The sequence of maturation, as indicated by PET scans of glucose metabolism, correlates well with the behavioral development that we observe in children.

The human pattern of a rapid demand for glucose followed by a gradual decline also occurs in cats and rhesus monkeys. For example, the critical period in cats for the primary visual cortex is 3 weeks to 180 days, when puberty takes place. As in

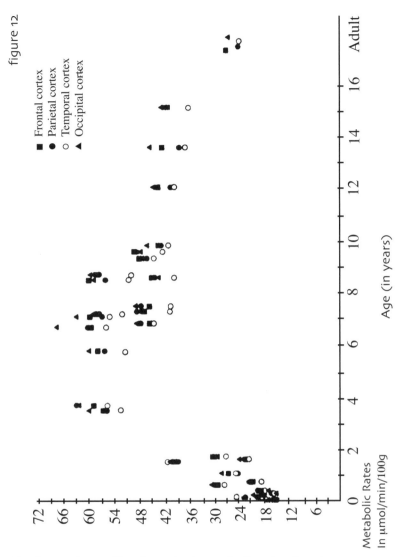

figure 12

Frontal cortex
Parietal cortex
Temporal cortex
Occipital cortex

Age (in years)

Metabolic Rates
In μmol/min/100g

Absolute values of local cerebral metabolic rates for glucose in cortical brain regions, plotted as a function of age in normal infants and children, and corresponding adult values.

Harry T Chugani (1997). Neuroimaging of developmental nonlinearity and developmental pathologies. In *Developmental Neuroimaging: Mapping the Development of Brain and Behavior*, RW Thatcher, GR Lyon, J Rumsey, and N Krasnegor, Eds. San Diego: Academic Press, p 190. Reprinted by permission of the author and of Academic Press.

human brains, the cat's brain wiring slows down in adolescence.

Examples of cognitive functions that are wired in the early years include language and ability to play a musical instrument. When a child learns a stringed instrument before the age of 12, the representation of the left hand (which plays the different notes) is expanded on the right cortex. Children learn a second language without accent and with fluent grammar until the age of 7 or 8. After that, we can still learn new languages, but it may be hard for us to fool you into thinking we're natives—language-learning ability usually declines even with a lot of practice. (However, we do know one person who came to the United States from Italy at the age of 15 knowing no English, and speaks flawlessly and without accent five decades later!)

Visual systems are subject to even more dramatic restrictions as they develop. Depriving an animal of normal visual experience during a short period after birth irreversibly changes neuronal connections. The animal may remain blind for life (Hubel and Wiesel, 1970). Although the neural basis of critical periods has been most studied in visual systems, similar phenomena occur in auditory, olfactory (pertaining to smell), and somatosensory systems as well.

Feral children (children raised in the wilderness without human parenting) learn language if they are captured before age 10, but not if they are captured later. Children who sustain brain injuries can usually recover up to age 10, because new connections continue to be laid down to compensate for damaged connections. An entire hemisphere can be removed before age 10 because of uncontrollable seizures, yet the child may acquire normal speech and a normal IQ. The ability to form new connections is called plasticity of the brain. Between ages 10 and 12, there is less recovery, and for adults there is even less. The window starts to close around age 10.

What do these findings mean for emotional development? Research on critical periods in emotional development is still limited. A few clues are available in studies of specific situations. For example, developmental specialists have wondered whether common fears in children and adults are the result of aversive conditioning (learned through unpleasant experiences), or innate "evolutionary-relevant" mechanisms for promoting survival. A particularly contentious debate has revolved around separation anxiety: do mothers or other environmental factors create it, or does it arise from children's inborn constitutions?

In a series of prospective longitudinal studies, Poulton and associates studied several common fears of heights, water, and separation from parents (Poulton, Milne, Craske, and Menzies, 2001). They found that fears of heights and water were primarily evolutionary-relevant, largely unrelated to a history of direct aversive conditioning. That is, these fears develop through natural selection to help the species avoid death by falling or drowning.

In order to answer the question with respect to separation anxiety, Poulton and his colleagues needed to ascertain whether known mechanisms for aversive learning—specific fear-generating events, modeling of fear by others, or transmission of frightening information—could account for separation anxiety at different developmental periods. They studied a variety of separation experiences from birth through age 18 (including loss of a parent through death or separation) and related these to measures of separation anxiety at ages 3, 11, and 18.

Certain planned separations in early to mid-childhood were modestly associated with lower levels of separation anxiety at later ages. Apparently the children got used (habituated) to being apart from their families through the experience of brief separations. Events at age 3 were not related to later separation anxiety, but at age 9, the child's vicarious learning (mother expressed fear of going out alone or being alone; teachers rated mothers as separation anxious or over-protective) showed a small relationship to separation anxiety at age 11. Although this "modeling" effect was the strongest evidence of learned separation anxiety, it accounted for only 1.8% of the variance in separation anxiety scores at age 11 (that is, 98.2% of the variance was due to other factors). Furthermore, it was not possible to tell whether the mother's fear was a modeling effect or a genetic effect. And the effect disappeared when the multiple regression equation was adjusted for the number of variables.

Death of a parent between the age of 11 and 18 and lower socioeconomic status were positively associated with separation anxiety at age 18, but each of these variables contributed only about 1% of the variance.

The investigators expected that children's hospitalizations would contribute to separation anxiety, because that kind of separation is unpredictable and often entails pain. However, just the reverse was true. Child hospitalizations before the age of 9 correlated inversely with separation anxiety at age 18. That is, more overnight stays in hospitals during childhood were related to less separation anxiety at age 18, suggesting an "inoculation" effect. At ages 11 and 18, females reported more separation anxiety symptoms than males, a finding consistent with previous research. However, given the very small percentage of variance attributable to any of the predictors, it was not surprising that no variables remained significant when adjustment was made for the number of variables in the equation.

Overall, the findings were consistent with the "nonassociative" theory of fear acquisition; separation anxiety does not arise from associations with childhood experiences but instead is inborn and is an evolutionary-relevant fear. However, even though separation anxiety is inborn, it can be modified by experiences such as planned separations or hospitalizations (reducing the level of separation anxiety), or by death of a parent or low socioeconomic status (increasing the level of separation anxiety).

In a study of cognitive development among infants with low birth weight, the relationship between zinc deficiency and cognitive development was explored (Bhatnagar and Taneja, 2001). Although we don't know the exact mechanisms by which zinc acts, the premise of the study was that zinc is essential for neurogenesis (formation of neurons), neuronal migration (travel from the neural tube to their final sites in the brain during gestation), and synaptogenesis (formation of synapses). Studies in animals have shown that zinc deficiency during the time of rapid brain growth appears to decrease cognitive activity, increase emotional behavior, and impair memory and the capacity to learn. Low maternal intakes of zinc during pregnancy and lactation were associated with less focused attention in newborns and decreased motor function at 6 months of age. Zinc supplementation resulted in better motor development and more playfulness in infants with low birth weight, and increased vigor and activity in infants and toddlers.

As I write, knowledge about critical periods in human emotional development is in its infancy. Research using imaging with children is restricted because PET and SPECT scans use radioactive substances, and fMRI, although not involving radioactivity, is not feasible for small children who cannot hold still for long periods of time. A well-developed inventory of critical or sensitive periods in human development is part of the research agenda for developmental researchers in the future.

21
Affiliation and Attachment

Affiliation has been defined as prosocial behavior that brings individuals into contact, a process of connecting or associating oneself (Insel and Winslow, 1998; Merriam-Webster, 1988, p 61). Examples of affiliation are a grasping response or a separation cry in an infant, children playing house, neighbors chatting on the sidewalk, or two adults becoming lovers. Affiliative behaviors are mediated by neurochemical systems that transmit monoamines (serotonin and dopamine) and neuropeptides (oxytocin, vasopressin, and the endorphins) interacting with sex and adrenal hormones (see Table 2, p 77).

Affiliation is fully developed only in birds and mammals. Deficits in affiliation are prominent in autism and in some forms of schizophrenia, manifested by a decrease in separation distress, lack of reciprocal social interaction, and inability to form social bonds. Knowledge about the neurochemistry of affiliation is therefore important for finding routes to treat these disorders, apart from interest in affiliation as a defining feature of humanness.

"A potent peptide prompts an urge to cuddle," proclaimed the New York Times (Angier, 1991, p. C1). This usually staid publication went on to rhapsodize about the neurotransmitter oxytocin, which doubles as a hormone. "New studies show an old hormone orchestrates many of life's pleasures and social interactions. . . . Just the potion for a bellicose world," the report continued. "In some cases, it acts as an aphrodisiac, inspiring males to seek females more ardently and females to invite their overtures more passionately. The hormone . . . helps stimulate the sensations of sexual arousal and climax. And after copulation, it acts as the proverbial cigarette, fostering a feeling of relaxed satisfaction. It makes new mothers more likely to nurture their young and new fathers happier to help out around the nest. Even among those who are neither sex partners nor parents, the compound can spur an overwhelming urge to cuddle" (ibid).

Since this glowing endorsement marked the popular debut of oxytocin 15 years ago, knowledge about affiliation has advanced. Because it's very difficult to impose experimental conditions on humans, most of the studies of connections between affiliative behaviors and neurochemistry have been carried out on animals. We'll look at a few of these and also at two experiments that were carried out with humans.

Grooming another individual is an important way that many animals, notably primates, affiliate. In fact, in many species this may be the predominant mode of socialization (Insel and Winslow, 1998; Dunbar, 1997). Although it may appear to human observers that a monkey energetically checking and combing another monkey's fur is trying to remove lice, grooming others is an important way of establishing alliances and reinforcing the dominance hierarchy, whether or not the groomed animal is infested. In fact, Dunbar (1997) suggests that this behavior is so important for social cohesion among primates that it represents a form of tactile "gossip"–that is, grooming may be the ape equivalent of conversations between customers and hairdressers at a salon!

Although grooming behavior appears simple, it's actually a complex social behavior with many possible different meanings. The social meaning of grooming varies from species to species and is associated with social status and kinship. The meaning of grooming varies, depending on how much time is spent, who receives it, relationship between groomed and groomer, and social status. In some species, dominants groom subordinates, in other species the reverse is true.

In both humans and primates, levels of serotonin metabolites (5-HIAA) in cerebrospinal fluid differ between individuals. These differences are at least partly genetic in origin. Serotonin metabolite levels have been positively correlated with time spent grooming and time spent in close proximity to other members of the social group and have been negatively correlated with aggression and impaired impulse control (Mehlman, Higley, Faucher et al, 1995). That is, the more 5-HIAA animals have, the more sociable and the less aggressive they are.

In a fascinating study of the relationship between neurotransmitters and social behavior, Raleigh and colleagues (1991) removed the dominant male from each of 12 social groups of vervets (small African monkeys) and treated one of the remaining males with a drug that was either serotonin-enhancing or serotonin-reducing. The design was a crossover study (that is, each treated male received each drug in two separate time periods). In every instance, the male who had received the serotonin-enhancing drug became dominant, but when the same male received the serotonin-reducing drug, another male in the group became dominant. Of particular interest to us is that the males who achieved dominance didn't do it by aggression. Instead, they became influential in their groups by increasing affiliative behaviors, including grooming, with females.

Hormones are clearly important in affiliation. Mastripieri

and Zehr (1998) administered the female sex hormone estrogen to female macaque monkeys whose ovaries had been removed. In response, the monkeys increased their handling of other monkeys' infants. Bardi and colleagues report association of female sex hormones and also cortisol, an adrenal hormone, with infant-directed behaviors in baboon mothers living in groups. Mothers who displayed more stress-related behaviors maintained less contact with their infants, and had higher postpartum cortisol levels and lower postpartum female sex hormones levels, than did other mothers. Interestingly, mothers overall who had higher prepartum cortisol levels showed higher levels of infant-directed affiliative behaviors after their babies were born. Their cortisol levels dropped dramatically soon after the infant's birth (Bardi, French, Ramirez, and Brent, 2004; Bardi, Shimizu, Barrett, et al, 2003) (See Chapter 19 for a discussion of cortisol).

The reasons for these findings are not clear, but Bardi and colleagues note the importance of the whole endocrine (hormone) system as a functional unit with respect to enhancing maternal care in primates. It appears that the dramatic physiologic changes from pre- to postpartum serve to help the mother cope with the challenges posed by the new infant. High levels of cortisol during pregnancy may be functional by ensuring that new mothers are aroused and attentive to their newborns. However, if the high levels persist for more than a brief period after birth, the negative effects described in Chapter 19 set in.

Diet can affect patterns of social and aggressive behavior of monkeys living in stable social groups (Simon, Kaplan, Hu, et al, 2004). Diets rich in soy protein and other soy products were fed to 44 adult male macaque monkeys over a 15-month period. The soy-fed monkeys engaged in intense aggression 67% more often and submissive behavior 203% more often than monkeys receiving a usual monkey diet. The soy-fed monkeys also spent 68% less time in physical contact with other monkeys, 50% less time near other monkeys, and 30% more time alone. The presumed reason for the marked effects of soy on affiliative and aggressive behavior was that soy-based foods bind with estrogen receptors and interact with estrogen-mediated responses. Estrogen produced by the processing of androgen (male hormone) facilitates typical male aggressive behavior. The decrease in affiliative behavior was not explained, but the well-documented strong association of serotonin with affiliative behavior suggested that this transmitter interacts with hormones.

The effects of environmental enrichment were studied in male mice to ascertain the relative roles of the physical and social components (Pietropaolo, Branchi, Cirulli et al, 2004). Mice 5 days old were randomly assigned to one of four different housing conditions for a period of 5 days: alone in a cage, with one other mouse in a cage, alone in a physically enriched cage, and with one other mouse in a physically enriched cage. As adults, 80 days after the exposure to physical enrichment environments, all mice were released into an open field.

In the open, mice who had been housed with companions in the physically enriched environment showed more affiliative and less aggressive social interaction and more frequently became dominant. They also showed changes in growth hormone levels. The combined physical and social enrichment appeared to increase brain plasticity (capacity for change) as well as the animals' ability to cope with social challenges.

In a novel experiment with human volunteers, Tse and Bond (2002) tested the effect of a selective serotonin reuptake inhibitor (SSRI), citalopram, on social behavior with an apartment mate and a stranger. Healthy volunteers (10 pairs) took part in a randomized double-blind crossover study in which subjects were given the drug for 2 weeks and a placebo for 2 weeks. That is, neither the subjects nor the investigators knew who was receiving the placebo or who was receiving the medication at any period. In each pair, one person (the subject) took the treatment, while the other (the apartment mate) did not. On the last day of each of the two treatment periods, the subjects socially interacted with a confederate (helper of the experimenter) behaving as a responsive person in a stranger/subject social interaction situation. After the interaction, the subjects played a game with the confederate that measures cooperative behavior and communication.

The subjects on medication reduced the number of points they awarded themselves and sent more cooperative messages during the game. They also showed a dominant pattern of eye contact in the stranger/subject social interaction. They were rated as significantly less submissive by their apartment mates than when they were on placebo. The researchers concluded that the administration of a serotonin-enhancing drug can modify social status in different interactions and can increase affiliative behavior (Tse and Bond, 2002).

In another human experiment, 50 healthy men and women volunteers (no history of a psychiatric diagnosis, psychotropic medication, or substance abuse) were randomly assigned to receive either placebo or paroxetine, a selective serotonin reuptake inhibitor, for 4 weeks. Subjects worked in pairs to solve problems collaboratively, with new partners on each trial, one in each pair receiving placebo, the other medication. The study was blind (neither subjects nor experimenters knew who received the drug). Changes in cooperative and noncooperative behaviors were coded from the beginning to the end of the experiment. Subjects receiving the medication showed a decrease in negative affect and an increase in affiliative and cooperative behaviors as compared with subjects receiving placebo. The degree of change correlated with measures of subjects' blood levels of the SSRI (Knutson, Wolkowitz, Cole et al, 1998).

DeVries and colleagues reviewed literature suggesting that social support in humans and affiliative behavior in some animals can provide a buffer against stress and can have a positive impact on measures of health and well-being (DeVries, Glasper, and Detillion, 2003). They compared HPA-axis activity among individuals who maintained relationships through aggressive displays with individuals who engaged in high levels of prosocial behavior. The word "individual" is key here, because an individual can be a human, another primate, a rodent, or other type of animal. These differences can occur within many species, not just between species. There were numerous instances when social interaction dampened HPA-axis activity and improved health outcomes, but little information emerged about the underlying mechanisms through which social behavior can provide a buffer against stress-related disease. Data suggested that oxytocin, well known for promoting social behavior, is the physiological link between positive social interactions and suppression of HPA axis activity. Other studies suggested that increases in serotonin brought about by handling stressed rat pups may have prevented the increase of glucocorticoid receptors. Thus serotonin may interact with other body chemicals to affect the hippocampus, stress responsiveness, and adult behavior (Insel and Winslow, 1998).

Attachment. We've seen that affiliation can range from casual association to strong bonding between lovers and between mothers and children. That is, affiliation sometimes progresses to attachment.

Psychology is replete with definitions for attachment. Bowlby (1982) defined it as a child being "strongly disposed to seek proximity to and contact with a specific figure and to do so in certain situations, notably when he is frightened, tired, or ill" (p 371). The caregiver responds by providing "comfort for distress, emotional availability for support, warmth and care for nurturance, and protection from danger" (Zeanah and Fox, 2004, p 34). Infants become attached to caregivers with whom they have a good deal of interaction, as the child learns that this person can be depended upon to respond when needed. Children seem to develop a hierarchy of preferences for these attachment figures (Zeanah and Fox, 2004).

Kagan (1994, pp 50-51) has culled three signs of infant attachment to a caretaker from research literature: the infant is more easily placated by the adults who care for him than by those who do not, is less distressed by the unfamiliar in the presence of these adults, and approaches them for play or solace. That is, the word "attachment" is used to denote a special emotional relationship that an infant has acquired with caregivers.

Sroufe (1997) views attachment as a system of dyadic emotion regulation in which caregivers help infants manage emotional tension. Kochanska (2001), citing several scholars of attachment, notes that attachment implies relatively complex constructs examining aspects of competence or risk such as resilience, confidence, qualities of representation of self and others, functioning in peer and romantic relationships, behavior problems, and psychopathology. Clearly, there is no consensus about what attachment is.

Before these various definitions of attachment existed, Mary Ainsworth, a pioneer in the attachment field, devised and standardized a set of procedures to measure how secure a child's attachment is to the mother (Ainsworth, Blehar, Waters, and Wall, 1978). These procedures, called the Strange Situation, consist of observing children between the ages of 9 and 23 months in an unfamiliar laboratory setting, for 3-minute periods, under four conditions: alone with the mother, alone with a stranger, with both the mother and the stranger, or all alone.

Responses are measured when the mother leaves the room and again when she returns. The mother leaves the child once with the stranger, once alone, and returns each time several minutes later. The child's immediate response to the mother's leaving, and his behavior when she returns, are supposed to indicate the quality of the child's attachment. According to the Ainsworth model, the most securely attached children show mild protest when the mother leaves, and are easily placated by her when she returns. Other categories of children are considered insecurely attached. Avoidant children do not protest when the mother leaves and do not approach the mother when she returns. Resistant children become very upset when the mother leaves and resist her attempts to soothe them when she returns. Disorganized attachment, a category added later to include many children who did not fit into the Ainsworth model, is characterized by behavior suggesting momentary confusion, fear, or disorientation upon reunion with a caregiver (Boris,

Hinshaw-Fuselier, Smyke et al, 2004). Boris and colleagues could find no association between the disorganized attachment classification and an attachment disorder diagnosis, although studies have shown that children receiving the disorganized attachment classification are at long-term risk for other psychiatric disorders (Lyons-Ruth and Jacobvitz, 1999; van Ijzendoorn, Schuengel, and Bakersman-Karnenburg, 1999).

Kochanska (2001) used the Strange Situation and other standardized laboratory procedures to explore individual differences in emotionality among young children. She wanted to see how these differences might be related to different attachment histories. The children were examined longitudinally from infancy to 3 years at 9, 14, 22, and 33 months of age to assess emotional responding. She focused on the emotions of fear, anger, and joy. Temperament scholars Kagan (1994) and Rothbart, Ahadi, and Evans (2000) view future emotional functioning as arising from early differences in temperament.

Kochanska was interested learning whether future emotional functioning might be arising from early patterns of attachment. By comparing children's responses to structured situations that typically elicit fear, anger, or joy, Kochanska was able to identify certain trends in emotional development that differed among children according to their attachment status based on the Strange Situation procedure. Overall, her study confirmed previous findings of the stability of temperament through the 2nd and 3rd years of life.

Although the children in her study did show different paths of emotional development according to the attachment classification that they received, her data could not demonstrate that the underlying cause of these differences was attachment status, rather than other possible variables such as temperament that might have contributed both to the attachment status behaviors and to the children's ongoing development.

Are attachment security categories from the Strange Situation valid measures of a child's attachment to the mother? Comparisons between different scales for measuring attachment showed disagreement about the major patterns of attachment (Boris, Hinshaw-Fuselier, Smyke et al, 2004). Stability of measures through time has also been a problem. In early studies, about half of children observed at 12 months and then at 19 months changed classification (Kagan, 1994). More recently, revisions of original classifications have reported greater stability (Waters, Merrick, Treboux et al, 2000).

What about the unusual laboratory situation itself? Imagine a 1-year-old in a strange place left alone or with a stranger. What other factors beside the child's relationship with the mother might determine whether a child becomes very upset, mildly upset, or not upset at all? Temperamentally anxious, fearful children could be expected to become anxious in this unfamiliar situation. But isn't it possible that children with calm, happy temperaments, who aren't easily frightened by the unfamiliar, might fail to get upset? Might such babies actually have a close relationship with the mother but not cry when she leaves, and, not having been upset when she left, might they fail to approach her anxiously when she comes back? Yet these children are classified as insecurely attached, avoidant type.

At the opposite end of the temperament spectrum, 1-year-olds who are very vulnerable to anxiety typically do become extremely upset when a mother leaves. Because they are extremely distressed, they sometimes continue to sob and push the mother away when she returns. They are classified as insecurely attached, resistant type. Again, does refusal to be soothed indicate a failure of attachment? Or might these very tense children have difficulty shifting gear from a distraught state to relief and pleasure about the mother's return? The failure of the attachment explanation leaves several questions unanswered.

Kagan (1994), a skeptic about the validity of the Strange Situation procedures, reports that in both Japan and West Germany, two countries with different child-rearing practices, 1-year-olds who were classified as insecurely attached in the Strange Situation had been very irritable as newborns. And in the middle-class West German group, only one-third of the children met the criteria for securely attached. Almost half were classified as avoidant because they did not greet the mother when she returned. Yet three-fourths of American children are classified as securely attached. Are most German children insecurely attached to their mothers, or are the differences in behavior related to socialization guided by a cultural norm of self-control and emotional reserve rather than a lack of attachment? Kagan (1994) documents numerous instances of differences in behaviors of toddlers across cultures that would mistakenly be labeled insecure attachment using the Strange Situation procedure. The Strange Situation certainly measures differences in behavior, but it's not clear what these differences have to do with the quality of a child's attachment to the mother.

The diagnosis of reactive attachment disorder (RAD) (American Psychiatric Association, 2000) denotes a set of characteristics known to be associated with grossly pathogenic care. That is, by definition RAD is a disorder caused by the environment, in the same way that the diagnosis of posttraumatic stress disorder (PTSD) is defined by its association with an environmentally induced traumatic experience. Children meeting criteria for RAD show markedly disturbed, developmentally inappropriate social behavior either as extreme withdrawal and inhibition, or by its converse, indiscriminate sociability such as excessive familiarity with strangers (American Psychiatric Association, 2000). This diagnosis overlaps poorly with categories of attachment security.

What are we to conclude from this puzzling collection of studies and opinions? Clearly there's plenty of room for disagreement. Studies of interventions designed to treat attachment disorders are few, because of the lack of agreement about the definition of attachment disorder (Boris et al, 2004). Recent reports of untested and even dangerous interventions for purported attachment disorders indicate a need for caution in work with children believed to be suffering from failures of attachment (Mercer, 2001).

22
Consciousness: An Evolutionary Model

Consciousness is the guarantor of all we hold to be human and precious. Its permanent loss is considered equivalent to death, even if the body persists in its vital signs.

Gerald M Edelman, 2004

Who am I? What is my Self? What is the meaning of life? Consciousness is the context in which such questions are framed. These questions, we believe, are uniquely human; we doubt that other species pose such questions. If that's true, then what biological differences between us and other species could account for human consciousness?

For centuries, consciousness has fascinated philosophers, theologians, biologists, psychologists, and ordinary people. The concept has been widely debated, but there's still little consensus about what consciousness really is. Dictionary definitions don't reflect the concept's complexity nor the controversy that surrounds it. For example, Webster's designates it as "the quality or state of being aware, especially of something within oneself" (Merriam-Webster, 1988, p 279). Although this definition is intuitively appealing, scholarly literature on consciousness suggests that it's inadequate.

Philosopher John Searle defines consciousness as "inner, qualitative, subjective states and processes of sentience or awareness. . . . [It] begins when we wake in the morning from a dreamless sleep and continues until we fall asleep again, die, go into a coma, or otherwise become 'unconscious' " (2000, p 559). Examples are feeling pain, seeing an object, feeling anxious, solving a crossword puzzle, trying to remember a phone number, arguing about politics, or wishing to be somewhere else (ibid).

Some authors equate consciousness with being aware that we're conscious, others with sensing ourselves as actors, agents, doers (the technical expression is "a sense of agency"). Primary consciousness (the state of being mentally aware of things in the world, characteristic of many animal species) has been distinguished from higher-order consciousness, which involves the ability to be conscious of being conscious, to recognize one's own acts and affections, to have a social concept of the self, to re-create past experiences, and to form future intentions (Edelman, 2004).

Higher-order consciousness requires at least semantic ability (ability to assign meaning to a symbol). Linguistic ability requires mastery of a whole system of symbols and mastery of a grammar. Chimpanzees have semantic but not linguistic ability; they appear to have the ability to be conscious of being conscious. It is believed that only humans have self-concepts, because linguistic ability is a prerequisite (ibid).

Most scholars seem to agree that consciousness differs from other neurobiological phenomena in that it is inherently subjective—it is experienced in the first person. The essence of the person's experience of consciousness cannot be captured simply by measuring and describing the neurobiological correlates of that experience. Yet identifying these neurobiological characteristics is a part of any scientific understanding, assuming we acknowledge that this information cannot convey subjective experience.

In this regard, saying "I am conscious" is different from saying "I'm angry" or "I'm scared" (emotional states) or "I don't know how to solve this algebra problem" (cognitive condition). Those specific emotional or cognitive states can often be related to particular neurobiologic structures, chemicals, and systems such as the amygdala, adrenaline, or the HPA axis. Consciousness, by contrast, is a more global state that appears to involve several brain systems. For this reason, it has been very difficult to visualize, much less demonstrate empirically, just which brain structures and systems participate in consciousness.

Today, beliefs about consciousness fall into two broad categories: scientific ("of, relating to, or exhibiting, the methods or principles of science") (Merriam-Webster, 1988, p 1051), and metaphysical ("of, or relating to, the transcendent or to a reality beyond what is perceptible to the senses") (ibid p 746). The scientific view holds that consciousness occurs in the form of brain functions performed by brain structures, even if we cannot specify what those functions and structures are. The metaphysical view asserts that consciousness transcends scientific principles.

The metaphysical view is in the tradition of René Descartes, who addressed the mind/body problem in the 17th century. Every person was a res cogitans (thinking thing), a unified rational acting entity in control of a body. However, according to Descartes, this thinking director-of-the-body was not subject to the laws of physics. Rather, it was nonmaterial and nonphysical, like the soul (Haldane and Ross, 1975).

Few scientists today still accept the notion that a thinking mind can exist devoid of a physical basis. Some humanities scholars such as Searle and philosopher Daniel Dennett endorse the scientific view as emphatically as do leading neuroscientists. However, the dualistic, dichotomous view considered in Part I has had a pervasive influence on academia as well as popular culture. Since the purpose of this book is to explore knowledge that integrates the bio with psychological and social components rather than separating the three spheres, readers wishing to learn more about the dualistic metaphysical perspective will probably want to look beyond this book.

Nobel Laureate Gerald Edelman (2004) has devised a model for the neurobiology of consciousness, sparking animated dialogue both within and outside the scientific community. This chapter introduces Edelman's model, which he calls the theory of neuronal group selection (TNGS). There isn't sufficient space to explore several models of consciousness in this book, so we've chosen the model that seems to have most excited scientists and nonscientists alike in the early 21st century. (The following discussion assumes that you're familiar with material presented in Part II of this book).

Edelman first proposed TNGS in 1978 and has elaborated it since then. It appears to be the most comprehensive model to date, and has received support from numerous leading neuroscientists, including Antonio Damasio, Oliver Sacks (neurologist known to the public through his book and the movie *Awakenings*), and Francis Crick, one of the scientists awarded the Nobel Prize in 1962 for the discovery of the double helix (Rothstein, 2004; Crick and Koch, 2003). Kim Dawson (2004) has extended Edelman's work in the area of time as a facet of consciousness. These scholars of neuroscience agree that the model is consistent with most of the relevant scientific evidence and offers a plausible explanation for consciousness. Data that could validate the model in its entirety are not yet available, however, so at the present time the TNGS awaits full validation.

Edelman's theory of consciousness encompasses some fundamental questions. How can the brain produce coherent activities such as speech, work, play, and social interaction that take place in a conscious state? How is it possible that 100 billion or more different nerve cells, acting in different parts of the brain, at different speeds, and in compliance with multiple sets of rules and procedures governing interacting body systems—how, in the context of such complexity, can the brain impose order rather than being overrun by chaos? Neuroscientists have been contemplating this mystery for some time and have named it the *binding problem.* To understand the neural basis for consciousness, a solution to the binding problem is required.

Edelman proposes that the workings of consciousness, and the coherence of brain activity despite its complexity, lie in the brain's developmental history and innate characteristics; in its presumed capability to self-organize through a process he dubs reentry; in the actions of at least three major brain systems; and in a selection process consisting of three overlapping steps. I'll review these topics to familiarize readers with some of the central concepts of TNGS—but I'd be less than forthright if I were to imply that I understand them fully!

This chapter departs from the rest of the book in that it focuses on a theory rather than phenomena that have been demonstrated through research. Although the theory is consistent with neuroscience research over the past two decades, the actual mechanisms by which consciousness takes place have not been demonstrated empirically as of the time of writing. The theoretical material that follows is challenging but fascinating. It is at the cutting edge of neuroscience explorations in the early 21st century. Critics can argue that TNGS is not critical for social work practitioners to know, and I agree. It's included for readers who would like to expand their horizons and stimulate their thinking.

The brain's developmental history and innate characteristics. In the brain, where's the master plan, and where's the executive director for this master plan? It would certainly seem that there must be one. Did you know that every attribute of a falling red ball is processed in a different location in the brain? In the case of the ball, there is color (red), shape (round), texture (rubbery), and movement (falling). There can be up to 30 different locations simultaneously processing different characteristics of a perceived object, yet we see only a single entity! We have a perception of an integrated object, the falling-red-ball. Surely there must be a brain equivalent of an air traffic controller directing the operation of the master plan.

Wrong, says Edelman. The coherence and integration that seem so amazing are achieved by quite different mechanisms. The brain develops from an embryonic structure called a neural tube. Precursor cells to neurons and glia (the cells that provide support to neurons) move to their final destinations in the brain, making various layers and patterns. As these precursor cells in the developing fetus differentiate into neurons and migrate to their final destinations in the brain, many die. Stronger cells survive and create new connections; weaker cells die off. Edelman likens brain development to evolution of the species. In fact, he designates the mechanisms of brain development "Neural Darwinism." The brain's power comes from natural selection acting in complex environments over eons.

As in evolution of the species, there's no superordinate brain region or executive program binding the color, form, texture, and movement of an object into a coherent perception.

Structures and functions in the brain respond at any given moment in time to the demands of the situation at that time–not according to a blueprint drawn by a long-term planning committee, and not overseen by a president's cabinet of ministers. In the process of evolution, species do not arise according to a preset program, and neither do our brains. Unlike computers, which execute programs that consist of logical steps following specific rules and procedures (algorithms), brains evolve through natural selection. Brains of higher-level animals construct patterned responses to events that are full of novelty. This behavior is unlike that of computers, which follow formal rules accompanied by explicit unambiguous instructions.

The brain is characterized by *variation*, *versatility*, and *degeneracy*. The concepts of variation and versatility are relatively easy to understand. We have many different kinds of neurons (at least 200) and many more brain cells than we need. So our brains are rich in numbers of cells and connections, variety, and patterns of cell movement.

Different individuals have different genetic influences, different epigenetic sequences (see Chapter 7), different bodily responses, and different histories in varying environments. The result is enormous variation at the levels of neuronal chemistry, network structure, synaptic strengths, timing of responses, memories, and motivational patterns. This variation endows the brain with versatility. The brain can choose which neuron or circuit will best respond to any unforeseen (environmental) input, and activate that structure.

This incredible diversity of material from which to select means that there's a lot of waste—cells and connections that end up not being used. However, this extravagance gives the brain remarkable resilience and potential for evolution. No brain event ever happens the same way twice. The astonishing variation in the human brain has created its reputation as the most complex entity in the known universe.

That's where degeneracy comes in. Most of us think of degeneracy in its common usage, defined as a state "characterized by deterioration" or the condition of a "morally degraded person" (Morris, 1975, p 347). But scientific meaning of the word is entirely different. A dictionary definition of the scientific concept of degeneracy is "taking on several discrete or distinct values or states" (ibid). Edelman's definition is "the ability of structurally different elements of a system to perform the same function or yield the same output" (2004, p 154).

Degeneracy accounts for the fact that when a certain gene is "knocked out" of a laboratory animal, 30% of the animals show no changes in the phenotype (observable characteristic) created by that gene. This is not because the gene wasn't really knocked out, but because degeneracy allows the brain to call up other different brain functions to produce the same effect.

As neurons develop, they form parallel groups, and they begin to fire in synchrony. This synchronous firing within bands of nerve fibers leads to hardwiring (neurons that fire together, wire together). These bands of fibers (sometimes called *nuclei*, the plural of nucleus) are thought to make up the larger systems within the brain that work together to create the phenomenon of consciousness.

How does this happen? Once neurons have migrated to their final destinations in the brain, they grow dendrites and axons, growth that is guided by physical and chemical factors (for a detailed explanation of early brain development, see Carlson, 2001, pp 73-76).

Axons form branches at their end. Each branch finds a vacant place on the membrane of an "appropriate" postsynaptic cell, grows a terminal button, and establishes a synaptic connection. "Appropriate" refers to a particular type of postsynaptic neuron, or even a different part of a neuron, appearing to secrete a specific chemical that attracts specific chemicals of axons (Carlson, 2001, p 75). The chemical signals that the cells exchange to tell one another to establish synaptic connections are currently being discovered by researchers.

When a presynaptic neuron establishes a connection, it receives a signal from the postsynaptic cell that allows it to survive. About 50% of neurons do not succeed in finding vacant spaces of the right type on neurons that they can connect to, and they die.

Once synaptic connections have been made, they can either become stronger or weaker. If a presynaptic neuron fires before an adjacent postsynaptic neuron fires, the connection—that is, the synapse—becomes stronger. If the time sequence is reversed, that is, if the postsynaptic neuron fires before the presynaptic neuron, the connection becomes weaker. So some connections become stronger and thrive, others become weaker. Connections that continue to weaken eventually die.

The phase of neuronal group formation is crucial to solving the binding problem. During the early establishment of brain structures, sets of interconnected neurons are existing side by side, but don't have much to do with each other—they happen to be hanging out, but not interacting. Variations in the patterns of connections among growing neurons form repertoires in each brain area composed of millions of variant circuits or neuronal groups. These variations arise at the level of synapses, because neurons that have fired together during embryonic and fetal stages of development have wired together.

When moments of synchrony between neurons begin to occur by accident, connections are made, and the more often these events are repeated, the stronger they become. Populations of synapses can become stronger or weaker. Here's how Dr Edelman explained it to me (personal communication, 2004).

Neurons are like musicians in a string quartet. Three of the four are playing a classical quartet from the eighteenth century (say, Haydn), while the fourth member, at the same time, is playing something hip and modern. Every so often, by chance, the deviant player and the Haydn players happen to hit an accented beat at the same moment. These occasional moments of synchrony form a connection. Both sides begin to play a bit more like the other, and the longer they play together, the stronger their connections and the more they play in sync. By the dress rehearsal, they are playing in almost perfect synchrony, but it's a new composition that integrates Haydn, Rock, and Rap.

Because the brain, unlike computers, is not logical, its central organizing principle is reentry. Reentry governs the space and time coordination among multiple networks in the brain. For example, the components of the falling red ball (movement, color, shape) are coordinated by the different brain regions involved communicating directly with each other through reentry.

Let's summarize principles central to the theory of neuronal group selection (TNGS) that we've considered so far.

1) *The brain was not created by design.* There was no master plan, and there is no overseer in the brain setting rules and making connections. The brain's power comes from natural selection acting in complex environments over eons of time. Like individuals who compete with each other for survival, brain cells also compete with each other for survival.

2) *The brain has an extraordinary amount of variation.*

3) *The brain is characterized by degeneracy, not efficiency or parsimony.* In the brain there is always more going on than seems necessary, more randomness, and more variation than in any system designed by humans. Degeneracy allows parts of the brain to "choose" other parts to connect to, flexibly responding through epigenetic mechanisms to demands created by input from the environment (see Chapter 7).

4) *Parallel group formation is one of the keys to the functions of neural systems* that allow different, separate neural systems to communicate with each other and to function in synchrony.

Reentry is pivotal to the TNGS. Reentry is a dynamic ongoing process of signaling back and forth across parallel fibers. These fibers are reciprocal, that is, they go in both directions between one neuron system and another. This process results in binding and is the basis for the emergence of consciousness. Reentry allows coherent, synchronized events to occur in the brain

(Edelman, 2004, p 174).

Reentry coordinates the activities of different brain areas in space and time. It binds neuron activities that are taking place at distances away from each other into circuits capable of coherent output. That is, reentry solves the binding problem.

Through reentry, color, orientation, and movement of a visual object can be integrated. No overarching map is necessary to coordinate these activities. These attributes communicate directly with each other through reentry, and are thus able to coordinate their messages so that we perceive a single entity in one time and one place.

Actions of three major brain systems. I stated earlier that three principal neural systems come into play, according to Edelman, in the creation of consciousness. Consciousness is not a property of a single brain location or neuronal type, but the result of dynamic interactions among widely distributed groups of neurons. Figure 13 presents sketches of these three systems.

1) *The thalamocortical circuits* (figure 13a) consist of tightly connected groups of neurons reaching both locally and across distances to form rich reciprocal connections between the thalamus and the cortex. The thalamus governs the levels of conscious states by regulating the amount of cortical activity. The content of conscious activity depends on which of the various cortical areas are active—for example, is it motor, visual, auditory, tactile, or some combination?

The thalamus is in the center of brain and is essential for conscious function, even though it's only somewhat larger than the last bone in our thumbs. All nerves from sensory receptors connect to the thalamus when they travel to the brain, through specific clusters of neurons called nuclei. These nuclei should not be confused with the generic concept of the nucleus of cells.

Postsynaptic neurons in each specific thalamic nucleus then project to axons that connect to particular areas of the cortex. The cortex receives messages from the thalamus, but there are also reciprocal axonal fibers that go back from the cortex to the thalamus. These reciprocal connections abound within the cortex.

Specific thalamic nuclei do not connect directly with any of the others. However, their activity is coordinated by a structure surrounding the thalamus called the reticular nucleus, which connects to the specific nuclei and which can inhibit their activity. The reticular nucleus acts as a switch or gate controlling the level of transmission of sensory messages. It "decides" how to balance nucleus A's level of activity with nucleus B's level of activity, perhaps allowing one to increase while holding the other constant. Another set of thalamic nuclei called the intralaminar nuclei is suspected of being essential for consciousness in that it sets appropriate thresholds or levels of cortical response.

2) The *basal ganglia* (figure 13b) are composed of loops of circuits whose neurons produce the inhibitory neurotransmitter GABA. These GABA neurons are in series with each other within the loops, so inhibitory GABA molecules are dispatched from the axon of one GABA neuron and picked up by the dendrites of the next GABA neuron in line. The action of those transmitters is to inhibit the action of the second GABA neuron.

The second GABA neuron now dispatches *less* GABA to the third neuron in the chain. That means it inhibits the third neuron less, so the third neuron becomes more excited. Thus we see that inhibitory neurons and neurotransmitters not only inhibit other neurons directly, but can also indirectly excite other neurons one step removed from them. That's the double negative effect we discuss in more detail in Chapter 29 on drug actions.

3) *Value systems* are widely spread out ascending projections of several systems, each of which processes a specific neurotransmitter (figure 13c). Value systems govern the threshold of activity of other neural structures. They generate motivation to survive and behaviors to carry out survival-directed activities such as the four Fs: feeding, fighting, fleeing, and

figure 13a

figure 13b

figure 13c

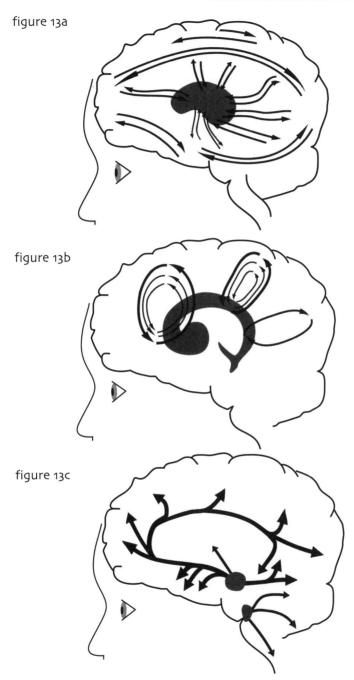

Adapted from Gerald Edelman, 2004, *Wider than the Sky*, Yale University Press, Figure 3, page 27. Printed with permission.

fornicating. The use of the word "value" here appears to mean "priority." The agendas of these circuits take priority over other possible brain actions that might not be crucial to survival. We saw sketches of the two value systems that produce and transmit dopamine and serotonin in Part II, figures 6a and 6b, pp 86-87 .

Value systems of neurons originate in lower regions of the brain, travel upward throughout almost all brain regions, spread out, and communicate widely throughout the brain. They behave like leaky garden hoses, spraying neurotransmitters into the far reaches of the brain. Each system of neurons specializes in a specific neuron and carries out various functions around the brain that utilize that specific transmitter. These systems specialize in dopamine, serotonin, norepinephrine, and several other transmitters. In order to promote the organism's survival, these value systems confer constraints on what other systems can do.

The many bundles of nerve fibers that form circuits in the thalamocortical circuits, the basal ganglia, and the value systems share information without going to any command-and-control center, overseer, or stage director. These circuits often travel in parallel routes, on paths within the brain that may be circular or unidirectional. Often they pass through gating stations that regulate their output, such as the reticular nucleus. This is reentry, the means by which they coordinate their activities.

The selection process that comprises the TNGS is made up of three overlapping steps.

1) *Developmental selection* during early brain formation involves epigenetic variations (see Chapter 7). As neurons form connections, patterns are formed in each brain area consisting of millions of variant circuits or neuronal groups. The variations arise at the level of synapses. According to Edelman, during embryonic and fetal stages of development neurons begin to fire together and as they do, they wire together.

2) *Experiential selection* overlaps with the first phase of

selection and also continues after major neuroanatomy is built. It creates large variations in synaptic strengths resulting from variations in environmental input. These synaptic modifications are subject to constraints of the value systems referred to above, systems of neurons geared to promotion of the individual's survival.

3) *Reentry.* During development, large numbers of reciprocal connections are established both locally and over long distances. These reciprocal neurons transmit signals around the circuits that make up the three major brain systems involved in consciousness: the thalamocortical circuits, the basal ganglia, and the value systems, depicted in Figure 13.

Are you still wondering how all of this works to make possible the phenomenon of human consciousness? So am I. Perhaps by the time the next edition of this book is written, it will be possible to fill in more details that can help us learn more about this amazing process.

References Part III

Alves SE, Akbari HM, Anderson GM, Azmitia EC, McEwen BC, and Strand FL (1997). Neonatal ACTH administration elicits long-term changes in forebrain monoamine innervation. Subsequent disruptions in hypothalamic-pituitary-adrenal and gonadal function. Annals of the New York Academy of Sciences 814:226-251.

Angier N (1991). A potent peptide prompts an urge to cuddle. New York Times, Jan 22, C1.

Barbazanges A, Piazza PV, LeMoal M, and Maccari S (1996). Maternal glucocorticoid secretion mediates long-term effects of prenatal stress. Journal of Neuroscience 16(12):3493-3549.

Bardi M, French JA, Ramirez SM, Brent (2004). The role of the endocrine system in baboon maternal behavior. Biological Psychiatry 55(7):724-32.

Bardi M, Shimizu K, Barrett GM, Huffman MA, Borgognini-Tarli SM.Bell M (2003).Differences in the endocrine and behavioral profiles during the peripartum period in macaques. Physiology and Behavior 80(2-3):185-94.

Bhatnagar S and Taneja S (2001). Zinc and cognitive development. British Journal of Nutrition 85 Suppl 2:S139-45.

Black JS (1998). How a child builds its brain: Some lessons from animal studies in neural plasticity. Preventive Medicine 27:168-171

Booth A, Johnson DR, Granger DA, Crouter AC, and McHale S (2003). Testosterone and child and adolescent adjustment: the moderating role of parent-child relationships. Developmental Psychopathology 39(1):85-98.

Boris NW, Hinshaw-Fuselier SS, Smyke AT, Scheeringa MS, Heller SS, Zeanah CH (2004). Comparing criteria for attachment disorders: establishing reliability and validity in high-risk samples. Journal of the American Academy of Child and Adolescent Psychiatry 43(5):568-77.

Brotman LM, Gouley KK, Klein RG, Castellanos FX, and Pine DS (2003). Children, stress, and context: integrating basic, clinical, and experimental research. Child Development 74(4):1053-57.

Caldji C, Diorio J and Meaney MJ (2000). Variations in maternal care in infancy regulate the development of stress reactivity. Biological Psychiatry 48(12):1164-1174.

Carlson NR (2001). Physiology of Behavior, 7th ed. Needham Heights, MA: Allyn and Bacon.

Chen X, Hastings PD, Rubin KH, Chen H, Cen G, and Stewart SL (1998). Child-rearing attitudes and behavioral inhibition in Chinese and Canadian toddlers: a cross-cultural study. Developmental Psychology 34:677-686.

Chugani HT (1997). Neuroimaging of developmental nonlinearity and developmental pathologies. In Developmental Neuroimaging: Mapping the Development of Brain and Behavior, RW Thatcher, GR Lyon, J Rumsey, and N Krasnegor, Eds. San Diego: Academic Press, pp. 187-195.

Chugani HT (1998). Unpublished data, reported in Talbot, 1998, p. 30.

Chugani HT, Phelps ME, and Mazziotta JC (1987). Positron emission tomography study of human brain functional development. Annals of Neurology 22:487-497.

Cintra, A, Bhatnagar M, Chadi G, Tinner B, Lindberg J, Gustafsson JA, Agnati LF, and Fuxe K. (1994). Glial and neuronal glucocorticoid receptor immunoreactive cell populations in developing, adult, and aging brain. Annals of the New York Academy of Sciences 746:42-61.

Courchesne E, Chisum H, and Townsend J (1994). Neural activity-dependent brain changes in development: Implications for psychopathology. Development and Psychopathology 6(4):697-722.

Crick F and Koch C (2003). A framework for consciousness. Nature Neuroscience 6(2):119-126.

Crockenberg SC (2003). Rescuing the baby from the bathwater: how gender and temperament (may) influence how child care affects child development. Child Development 74(4):1034.

Dawson KA (2004). Temporal organization of the brain: Neurocognitive mechanisms and clinical implications. Brain and Cognition 54:75-94.

Dennett DC (2003). The self as a responding–and responsible–artifact. Annals of the New York Academy of Sciences 1001:39-50.

Dennett DC (1995). Review of Antonio R Damasio, Descartes' Error: Emotion, Reason, and the Human Brain. New York Times Literary Supplement, Aug 25 pp 3-4.

Diego MA, Field T, Hernandez-Reif M, Cullen C, Schanberg S, Kuhn C (2004). Prepartum, postpartum, and chronic depression effects on new borns. Psychiatry.67(1):63-80.

Dunbar RIM (1997). Grooming, Gossip, and the Evolution of Language. Cambridge, MA: Harvard University Press.

Edelman GM (2004). Wider than the Sky. New Haven: Yale University Press.

Escorihuela RM, Tobena A, and Fernandez-Teruel A (1994). Environmental enrichment reverses the detrimental action of early inconsistent stimulation and increases the beneficial effects of postnatal handling on shut tlebox learning in adult rats. Behavioural Brain Research 61(2):169-173.

Field T (1995). Massage therapy for infants and children. Journal of Developmental and Behavioral Pediatrics 16(2):105-111.

Fisher PA and Chamberlain P (2000) Multidimensional treatment foster care: a program for intensive parenting, family support, and skill-building. Journal of Emotional and Behavioral Disorders 8:155-164.

Fisher PA, Gunnar MR, Chamberlain P, and Reid JB (2000). Preventive intervention for maltreated preschool children: Impact on children's behavior, neuroendocrine activity, and foster parent functioning. Journal of the American Academy of Child and Adolescent Psychiatry 39:1356-1364.

Fisher PA, Ellis BH, and Chamberlain P (1999). Early intervention foster care: A model for preventing risk in young children who have been maltreated. Children Services: Social Policy, Research, and Practice 2(3):159-182.

Fox NA (1991). If it's not left, it's right. American Psychologist 46:863-72.

Fuller L and Clark LD (1968). Genotype and behavioral vulnerability to isolation in dogs. Journal of Comparative and Physiological Psychology 66: 151-156.

Fuxe K, Diaz R, Cintra A, Bhatnagar M, Tinner B, Gustafsson JA, Ogren SO, and Agnati LF (1996). On the role of glucocorticoid receptors in brain plasticity. Cellular and Molecular Neurobiology 16(2):239-258.

Gould E, Reeves AJ, Graziano MS, and Gross CG (1999). Neurogenesis in the neocortex of primates. Science 286:548-552.

Granger DA and Kivlighan KT (2003). Integrating biological, behavioral, and social levels of analysis in early child development: progress, problems, and prospects. Child Development 74(4):1058-63.

Groza V, Ryan SD, and Cash SJ (2003). Institutionalization, behavior, and international adoption: predictors of behavior problems. Journal of Immigration and Health 5(1):5-17.

Haldane E and Ross G, eds (1975). The Philosophical Works of Descartes. Cambridge, UK: Cambridge University Press.

Hameroff S and Penrose R (1996). Conscious events as orchestrated space-time selections. Journal of Consciousness Studies 2:36-53.

Hubel DH and Wiesel TN (1970). The period of susceptibility to the physiological effects of unilateral eye closure in kittens. Journal of Physiology 206:419-436.

Insel TR and Winslow JT (1998). Serotonin and neuropeptides in affiliative behaviors. Biological Psychiatry 44(3):207-219.

Insel TR and Hulihan TJ (1995). A gender specific mechanism for pair bonding: Oxytocin and partner preference formation in monogamous voles. Behavioral Neuroscience 109:782-89.

Johnson JE, Christie JF, and Yawkey TD (1999). Play and early childhood development, 2nd ed. New York: Longman.

Jones NA, McFall BA, Diego MA (2004). Patterns of brain electrical activity in infants of depressed mothers who breastfeed and bottle feed: the mediating role of infant temperament. Biological Psychology 67(1-2):103-24.

Kagan J (2003). Biology, context, and developmental inquiry. Annual Review of Psychology 54:1-23.

Kagan J (1994). The Nature of the Child (10th Anniversary edition). New York: Basic Books.

Kagan J, Snidman N, and Arcus D (1998). Childhood derivatives of high and low reactivity in infancy. Child Development 69(6):1483-93.

Knutson B, Wolkowitz OM, Cole SW, Chan T, Moore EA, Johnson RC, Terpstra J, Turner RA, and Reus VI.(1998). Selective alteration of personality and social behavior by serotonergic intervention. American Journal of Psychiatry 155:373-379.

Kochanska G (1993). Toward a synthesis of parental socialization and child temperament in early development of conscience. Child Development 64:325-347.

Kochanska G (2001). Emotional development in children with different attachment histories: the first three years. Child Development 72(2):474-90.

Liu D, Diorio J, Tannenbaum B, Caldji C, Francis D, and Freedman A (1997). Maternal care, hippocampal glucocorticoid receptors, and hypothalamic-pituitary-adrenal responses to stress. Science 277(5332):1659-1662.

Lorenz K (1952). King Solomon's Ring: New Light on Animal Ways. New York: Crowell.

Lyons-Ruth K and Jacobvitz D (1999). Attachment disorganization: unresolved loss, relational violence, and lapses in behaviora and attentional strategies. In J Cassidy and PR Shaver, eds, Handbook of Attachment: Theory, Research, and Clinical Applications, pp 520-554.

Masten AS, Hubbard JJ, Gest SD, Tellegen A, Garmezy N, and Ramiriez M (1999). Competence in the context of adversity: pathways to resilience and maladaptation from childhood to late adolescence. Developmental Psychopathology 11(1):143-69.

Mastripieri D and Zehr JL (1998). Maternal responsiveness increases during pregnancy and after estrogen treatment in macaques. Hormones and Behavior 34:223-230.

McManis MH, Kagan J, Snidman NC, and Woodward SA (2002). EEG asymmetry, power, and temperament in children. Developmental Psychobiology 41(2):169-77.

Meaney MJ (2001). Maternal care, gene expression, and the transmission of individual differences in stress reactivity across generations. Annual Review of Neuroscience 24: 1161-1192.

Meaney MJ, Diorio J, Francis D, Widdowson J, LaPlante P, Caldji C, Sharma S, Seckl JR, and Plotsky PM (1996). Early environmental regulation of forebrain glucocorticoid receptor gene expression: implications for adrenocortical responses to stress. Developmental Neuroscience 18(1-2):49-72.

Mehlman PT, Higley JD, Faucher I, Lilly AA, Taub DM, Vickers J et al (1995). Correlation of CSF 5-HIAA concentration with sociality and the timing of emigration in free-ranging primates. American Journal of Psychiatry 152:6.

Merriam-Webster (1988). Webster's Ninth New Collegiate Dictionary. Springfield, MA: Merriam Webster.

Morris W, ed. (1975). The American Heritage Dictionary of the English Language. Boston: Houghton Mifflin.

National Institute of Child Health and Human Development (2004). Are child developmental outcomes related to before- and after-school care arrangements? Results from the NICHD study of early child care. Child Development 75(1):280-95.

Nelson JD and Panksepp JB (1996). Oxytocin mediates acquisition of maternally associated odor preferences in preweanling rat pups. Behavioral Neuroscience 110:583-592.

Newcombe (2003). Some controls control too much. Child Development 74(4):1050-2.

Patchev VK, Montkowski A, Rouskova D, Koranyi L, Holsboer F, and Almeida OF (1997). Neonatal treatment of rats with the neuroactive steroid tetrahydrodeoxycorticosterone (THDOC) abolishes the behavioral and neuroendocrine consequences of adverse early life events. Journal of Clinical Investigation 99(5):962-966.

Pietropaolo S, Branchi I, Cirulli F, Chiarotti F, Aloe L, and Alleva E (2004). Long-term effects of the periadolescent environment on exploratory activity and aggressive behaviour in mice: social versus physical enrichment. Physiology and Behavior 81(3):443-53.

Poulton R, Milne BJ, Craske MG, and Menzies RG (2001). A longitudinal study of the etiology of separation anxiety. Behavior Research and Therapy 39(12):1395-1410.

Purves D, Augustine, Fitzpatrick GJ, Katz D, LaMantia LC, McNamara AS, JO, and Williams SM (2001).Neuroscience, 2nd ed.. Sunderland, MA: Sinauer Associates.

Purves WK, Sadava D, Orians GH, and Heller HC (2001). Life: The Science of Biology, 6th ed. Sunderland, MA: Sinauer Associates.

Rothbart MK 1989). Temperament and development. In GA Kohnstamm, JE Bates, and MK Rothbart, eds, Temperament in Childhood pp 187-247. Chichester, England: Wiley.

Raleigh MJ, McGuire MT, Brammer GL, Pollack DB, and Yuwiler A (1991). Serotonergic mechanisms promote dominance acquisition in adult male vervet monkeys. Brain Research 559(2):181-90.

Rothbart MK, Ahadi SA and Evans DE (2000). Temperament and personality: origins and outcomes. Journal of Personality and Social Psychology 78(1):122-135.

Rothstein E (2004). The brain? It's a jungle in there. New York Times Arts and Ideas, March 27, p B7.

Rutter M (1998). Developmental catch-up, and deficit, following adoption after severe global early privation. Journal of Child Psychology and Psychiatry 39(4):465-476.

Scafidi F and Field T (1996). Massage therapy improves behavior in neonates born to HIV-positive mothers. Journal of Pediatric Psychology 21(6):889-897.

Searle JR (2000). Consciousness. Annual Review of Neuroscience 23:557-578.

Selye H (1976). The Stress of Life. New York: McGraw-Hill.

Simon NG, Kaplan JR, Hu S, Register TC, and Adams MR (2004). Increased aggressive behavior and decreased affiliative behavior in adult male monkeys after long-term consumption of diets rich in soy protein and isoflavones. Hormones and Behavior 45(4):278-84.

Slotkin TA, Barnes GA, McCook EC, and Seidler FJ (1996). Programming of brainstem serotonin transporter development by prenatal glucocorticoids. Brain Research and Developmental Brain Research 93(1-2):155-61.

Sroufe LA (1997). Emotional Development. New York: Cambridge University Press.

Talbot M (1998). Attachment theory: the ultimate experiment. New York Times Magazine (May 24): 24-54.

Thomas A and Chess S (1977). Temperament and Development. New York: Brunner/Mazel.

Tse WS and Bond AJ (2002). Serotonergic intervention affects both social dominance and affiliative behaviour. Psychopharmacology (Berl) 161(3):324-30.

Van den Boom DC (1994). The influence of temperament and mothering on attachment and exploration: an experimental manipulation of sensitive responsiveness among lower-class mothers with irritable infants. Child Development 65:1457-77.

Volkmar FR and Greenough WT (1972). Rearing complexity affects branching of dendrites in the visual cortex of the rat. Science 176(42):1145-1147.

Waters E, Merrick S, Treboux D, Crowell J, and Albersheim L (1997). Attachment security in infancy and early adulthood: a twenty-year longitudinal study. Child Development 71(3):684-9.

Werner EE (1989). Children of the Garden Island. Scientific American 260(4):106-111.57

PART IV
MAJOR PSYCHIATRIC DISORDERS

PART IV
MAJOR PSYCHIATRIC DISORDERS:

In Part IV, we'll consider atypical psychological functioning that causes problems for individuals and/or others around them. These psychological characteristics are defined as psychiatric disorders when aspects of affect, behavior, or cognition are deemed abnormal and problematic. They are evaluated in relation to duration, severity, presence across social settings, situations in which they do or do not occur, context of the onset, social and economic stressors, cultural forces, and underlying neurobiological structures and functions.

Current evidence highlights several facets of contemporary mental health practice in light of knowledge about these conditions.

First, most major diagnostic groups using the *Diagnostic and Statistical Manual of Mental Disorders* classifications (APA, 2000) include different but related conditions. Examples of these diagnostic groups are unipolar depression, bipolar disorder, and schizophrenia, each of which includes more than one subcategory. People meet the criteria for a subcategory because they show certain types of affect, behavior, or cognition. However, most diagnoses allow for choices of a subset of characteristics (say, five) from among a much longer list (say, twelve) in order to qualify for the diagnosis. For this reason, there can be a considerable difference in the observable characteristics of one person meeting criteria for a diagnosis from those of another person meeting criteria for the same diagnosis.

A second aspect is that an individual's observable characteristics are associated with specific underlying neurobiological features (such as particular neurotransmitters, neural structures, and systems of neurons) (Pennington, 2002). Even when the observable behaviors are the same, the biological underpinnings may differ, as in the case of unipolar depressions that are and are not responsive to selective serotonin reuptake inhibitors (SSRIs). It's also important to remember that psychosocial phenomena affect neural chemistry on a continuous basis—when we experience psychosocial events, our brains respond with changes in chemistry. The psychotherapeutic process itself acts on neurobiology while it's acting to affect the psyche.

For any client, we ask ourselves how a particular psychosocial intervention or medication can help this person. The answer may depend on how a particular intervention affects the person's neurobiology. For any given diagnosis, different medications and different psychosocial interventions are likely to be effective. Furthermore, sometimes a medication can have the same neurobiological effect as a nonmedical intervention, even though the results were achieved by different routes.

For example, some symptoms of obsessive-compulsive disorder (OCD) can be treated with behavior therapy, medication, or a combination of both. Activity in the caudate nucleus is known to be related to symptoms of OCD. By measuring brain glucose metabolism using PET scans, researchers compared a group of people with OCD treated with behavior therapy to another group treated with medication (Schwartz, Stoessel, Baxter et al, 1996; see Chapter 4 for a discussion of imaging). The behavior therapy consisted of 10 weeks of exposure therapy and response prevention interventions (ibid; Kazdin, 2001; De Silva, Menzies, and Shafran, 2003). Behavior therapy responders had symptom remission and corresponding decreases in caudate glucose metabolic rates similar to those of medication responders.

Does this mean we might dispense with medication altogether in the treatment of psychiatric disorders and use only psychosocial interventions? Absolutely not—because this clear-cut similarity in action is the exception rather than the rule. For many conditions, medication is the only effective treatment we currently have at our disposal. For other conditions, combinations of medication and nonmedical interventions are required, with different contributions from each. What this OCD research illustrates is that when psychological therapies are effective, they are acting, invisibly, at the neurobiological level. Psychotherapy changes brain metabolism, and psychotherapy and medication can affect the same neural systems (Pennington, 2002).

A third aspect is that research in the 21st century is focusing increasingly on underlying neurobiological factors associated with a particular psychological characteristic or symptom. Mental health researchers are trying to learn about *pathophysiology* (abnormality in physiological process) that accompanies observable psychological challenges or problems. This information is essential for helping us find effective medications to alter underlying disordered processes. Since each component of the individual's physiological processes interacts with other components, the task is challenging, sometimes daunting.

Fourth, the mental health field is changing. Theory-based approaches are giving way to evidence-based interventions tailored to specific problems associated with underlying neurochemistries (Livesley, 2004; US Department of Health and Human Services, 1999). Evidence-based practice searches for answers to intervention-related questions. What's the most effective way to treat that particular *problem* (as opposed to that particular diagnosis) in these particular environmental circumstances? What mechanisms operate to bring about the desired change? How can different interventions be integrated into a comprehensive plan or treatment package (Livesley, 2004)?

Doesn't this development in the mental health field jeopardize a strengths perspective (see Chapter 5)? No, not at all, if we maintain awareness of strengths as well as problems and a commitment to strengths-based intervention. Holding a strengths perspective does not mean pretending that problems are not there. If our clients didn't have problems, they wouldn't need our services. Yet I fully endorse principles of strengths-based practice such as those presented by Dennis Saleebey (summarized in Chapter 5) and others.

An area in which some advocates of strengths-based practice may differ from me is in their understanding of the medical model. I strongly support learning as much as possible about the ways our bodies and brains enhance or impede our quality of life. This position requires basic biological knowledge, especially neuroscience information, that's central to issues in child development, mental health, mental illness, substance abuse, and addiction—that is, medical information in the context of an epidemiological, systems-oriented approach.

Evidence is already very strong, and rapidly increasing, that a systems understanding of all of these issues requires information about neurobiological underpinnings of the challenges we're trying to address (see Chapter 2 for a discussion of the reasons). Why should incorporating scientific information into assessment and intervention prevent workers from having a respectful attitude, humility about how little we really do know, and understanding that a person is not equivalent to the disability he or she has?

We do need to retrain ourselves to use "people first" language, e.g. a person with schizophrenia or a person addicted to cocaine rather than a schizophrenic or a cocaine addict. But why can't a respectful, collaborative posture based on the concept of worker-client partnership include exploring neurobiological aspects of psychological and addictive issues? Epidemiological and complex adaptive systems models assume that physiological interventions using medication or nondrug biological alternatives are protective factors that can make enormous, sometimes even lifesaving differences in people's lives (see Chapters 2, 5, and 7).

One objection that critics of medical models have advanced is its normative stance with regard to "pathology." Who are we to define certain groups of people as "pathological," others as "normal"? I agree with this critique.

Chapter 23 considers the cultural relativity of the concept of normality and problems with its use. We then consider the assessment of psychiatric disorders, its evolution across mental health disciplines, and contemporary multisystem assessment and intervention through the prism of complex adaptive systems (Chapter 24).

A practice example of the interplay of professional culture, organizational agendas, and the process of diagnosis follows, all in a situation where neuroscience knowledge is extremely important for assessment and intervention (Chapter 25). To illustrate ways that psychological and neurobiological characteristics connect, as well as ways that multisystem interventions can be responsive to biopsychosocial needs, we review research on the origins of and interventions for one diagnosis, borderline personality disorder (BPD) (Chapter 26). This analysis is an example of neuroscience information and evidence about interventions related to a particular psychiatric diagnosis that has emerged through recent research. Similar kinds of knowledge are essential to begin to understand, assess, and provide help for any psychiatric disorder.

Finally, we'll look at a detailed practice example of a complex adaptive systems approach to assessment and intervention planning (Chapter 27). The situation involves a young woman with characteristics of several DSM diagnoses, members of her family, and cultural issues affecting the family.

23
Normality, Professional Culture, and Psychiatric Disorders

What's "normal"? Does the answer to this question depend on whom you ask? Yes, to a certain extent. Dominant cultures and subcultures strongly influence a wide range of beliefs, values, and meanings in most areas of human society, including mental health and mental illness. The National Institute of Mental Health's Culture and Diagnosis Group defines culture as "meanings, values, and cultural norms that are learned and transmitted in the dominant society and within its social groups. Culture powerfully influences cognition, feelings, and self-concept *as well as the diagnostic process and treatment decisions*" (Sadock and Sadock, 2003, pp 168-169, emphasis added).

That is, concepts of pathology implicit in the notion of diagnosis are strongly influenced by a particular American subculture, that of the mental health professions such as psychiatry, clinical psychology, and clinical social work. (The meaning of the word "clinical" is beyond the scope of our discussion. The term is used here to refer to subsets within the disciplines of psychology and social work that designate themselves as clinical). Let's look at an example of the definition of illness versus health in two subcultures.

> Jared, age 27, is kneeling on his back porch talking audibly to someone whom no one else can see or hear. He does this several times a week. His next-door neighbor Beatrice is sitting on her back porch, which is almost identical to Jared's on this street of row houses. She scowls and mutters under her breath. "When is he going to see a shrink? He's a nut case."
>
> Jared's neighbor Edna on the other side, however, nods approvingly and crosses herself. "Praise the Lord, praise the Lord, he's found the Lord," she says to her sister as they both relax on their lawn chairs and enjoy the late afternoon sun. Edna and Jared's mother attend the same church.

Who's right? Is Jared mentally ill? Is he hallucinating? Does he have a diagnosis of schizophrenia? Or is he having a life-expanding spiritual experience?

In Jared's neighborhood, no one has ever taken any action against him although his behaviors are well known to everyone on the street. He is not violent, has never hurt or even frightened anyone, and he's lived there for years. Jared does not work. He stopped attending school when he was 13 because his mother didn't like the way the school was treating him. His mother supports them both by working in a local store and by part-time housekeeping at a motel. When she is not at work, they spend a lot of time visiting family members who live nearby. Jared has never received mental health treatment. Family members have never urged Jared's mother to "do something about Jared"—he is viewed as one of them and acceptable as he is.

One American subculture, that of psychiatry, would probably look at his behavior in light of DSM-IV criteria and find him to have schizophrenia, chronic undifferentiated type. Members of another subculture of American society, the religious group that Jared's mother and next-door neighbor Edna are part of, might interpret his behavior as a spiritual experience, not an illness. Probably some of the residents of the neighborhood would characterize Jared in a third way, as an odd or peculiar but harmless young man.

But let's consider what may happen when a new element is introduced into this history.

Jared attacked two little girls when they laughed at him and taunted him with words such as "you're crazy" and "you've got a loose screw." He flew into a rage, punched both girls and grabbed one of them. He banged her head against the ground until she became unconscious. She recovered with no appearance of long-term damage. Jared was apprehended, spent a year in jail, and was released to his mother's home with a mandate to receive court supervision and psychiatric treatment. He received a diagnosis of schizophrenia, chronic paranoid type. He was required to attend a day treatment program and to report to his parole officer regularly. Jared and his mother were warned that if Jared violated the conditions of his parole he would be sent to prison for a long time. Jared said he was sorry and wouldn't do anything like that again.

No further incidents have occurred, but Jared's mother is now on tenterhooks. She is afraid to leave him alone, and has had to reduce her work hours to be with him when day treatment is not open. Under these altered conditions, subcultures redefine the individual. Psychiatry defines him with a more severe diagnosis, and many in the local neighborhood, who formerly perceived him as odd or different, now see him as insane, violent, and dangerous to children. Neighborhood parents warn their children to stay away from Jared.

How did Jared's status change from being odd, but acceptable, to mentally ill, dangerous, and an ex-offender? The concepts "abnormal" and "pathological" are usually not invoked for all unusual psychological functions but specifically for those behaviors, emotions, or cognitions that cause trouble for the individual and/or the people who must interact with him or her. Had Jared remained the same, without the incident having occurred, it's likely he would never have been defined as having a specific mental illness because he would not have encountered the professional subculture that defines people on that dimension.

Scholars of mental disturbances have arrived at concepts of normality at least in part from studying behaviors, emotions, or cognitions that cause problems for the individuals themselves or for others, in such areas as interpersonal relations, school and work performance, and satisfaction in life. Where do we draw the line between normal and abnormal, mentally intact and mentally ill? Are suicide bombers mentally ill? Wouldn't an American who voluntarily dies in order to inflict death on others be considered mentally ill? Might committing suicide be a rational act, given its cultural and political context? What if this suicide bomber's cause is so compelling that he believes the price of death is worth it? What if his culture applauds death for that cause, and assures the individual of rewards in the hereafter?

The world of clinical diagnosis in the United States has developed a set of criteria for mental disorders that draws an arbitrary line between mental health and mental illness. However, its framers acknowledge the importance of culture in the definitions by allowing exclusion from the diagnosis when cultural differences appear to explain deviant behavior (APA, 2000).

Authors of the DSM-IV-TR and some major textbooks on psychiatric disorders stop short, however, of spelling out how their approach to assessment *reflects their own subculture*, that of professional mental health practitioners who have achieved certain status and income-generating capacity by learning about the kinds of problems that bring people to their services and about methods for addressing these problems (APA, 2000; Sadock and Sadock, 2003). I'm not suggesting that DSM diagnostic categories are necessarily inaccurate or irrelevant (some classifications are better validated than others). The issue of validity of the many diagnostic categories currently in use is beyond the scope of this book. What I am suggesting is that it would be desirable for these authors to add another component to their analysis: how professional culture contributes to their definitions of mental illnesses and their repertoire of treatments for these diagnoses.

24
Assessment and Intervention: Evolution Toward a Complex Adaptive Systems Perspective

What are we assessing when we begin work with a client, and as we continue work with the client? In order to understand how contemporary assessment has developed from assessment practices common in earlier decades, some elements of the history of social work assessment is needed. A history of practice ideologies was summarized in Chapter 2. *Diagnosis* and *treatment*, in the terminology of the 1960s and '70s, were the precursors of current approaches to *assessment* and *intervention*, terms that most social workers prefer today. Psychiatry and other disciplines working in medical settings continue to use the diagno-

Table 4A

General Kinds of Information Needed to Carry Out Assessments: Early and Recent Nonbiological Approaches Compared

Psychosocial Diagnosis: 1950s–1980s	Multisystem Assessment before Emergence of Neurobiology: 1970s–early 1990s
1. What is the client's presenting problem(s)? Client's view of the problem. View of the problem by significant others.	The same
2. What is the history of the presenting problem?	The same
3. What is the client's current life situation and family history (family, work and school, health issues in family, economic security, religious involvement, social interaction outside family)?	The same
4. How do members of the client's family interact with each other?	The same
5. How have previous helpers, if any, tried to alleviate the problem(s)? With what results?	The same

---

sis/treatment terminology, however researchers and practitioners in these fields are increasingly considering environmental factors in their evaluations.

In the 1940s, '50s, and early '60s, social work students in most MSW programs were trained to focus on intrapsychic

Table 4B

General Kinds of Information Needed to Carry Out Assessments:
Early and Recent Nonbiological Approaches Compared
(continued)

Psychosocial Diagnosis: 1950s-1980s	Multisystem Assessment before Emergence of Neurobiology: 1970s-early 1990s
6. What is the client's history of relationships withsignificant others in the family of origin?	6. How are stressors from environments outside the family impacting family members and influencing family dynamics?
7. What does the preceding information tell us about how the presenting problem has arisen from this history of relationships in the family of origin? (Here, interpretation is advanced based on assessment of ego function, early developmental conflicts and fixations, and other aspects of psychoanalytic developmental theory.)	7. How does cultural diversity influence family members? How do the worker's subcultures and the organizational subcultures of the agency providing services impact the client and family members?
	8. What environmental and individual risk factors and protective factors are associated with this problem?

---

process (what's happening inside the individual psyche). This process was inferred from the person's words and behaviors. Problematic relationships in the present were believed to be the end product of a developmental process starting from infancy. Adult ways of relating were assumed to be the legacy of early mother-child dynamics, embellished with contributions from relationships with other members of the nuclear family of origin, i.e. the father and siblings. Cultural differences were the subject of some academic courses, but were typically viewed as variations more in style than substance. The substance was psychic functioning in the context of present relationships. Whatever the age of the client, psychic functioning at that point in time was assumed to result from interpersonal influences within the nuclear family of origin.

Biological science of the era was not usually incorporated into the assessment other than to note that the individual client or a member of the client's family once had or was currently suffering from illnesses such as diabetes or cancer. Such illnesses were included in the assessment as situational factors, i.e. general sources of stress as opposed to specific biochemical processes that might affect the individual's psychological functioning. These illnesses were considered to be physical, in contrast to psychological illnesses which were mental. Freud and his followers connected biology to psychology mostly in relation to interpersonal transactions between mother and child that took place in the context of eating (oral phase), eliminating (anal phase), and early cognizance of one's sexual equipment (genital or phallic phase). The child's psychological development was presumed to depend on *how* the mother treated her child during each developmental phase. The mother's behavior, independent of the child's innate characteristics, was viewed as the decisive force in the child's mental health as an adult.

A typical clinical assessment of the day was modeled on medical diagnosis and was called a psychosocial diagnosis. It consisted of a statement of the presenting problem, a description of the onset and progression of the problem, precipitating factors that appeared to have sparked the onset, a developmental history of the individual in his or her family of origin, an interpretation of this information using the language of ego psychology, and recommendations for treatment. For example, the interpretation section of the psychosocial diagnosis emphasized the client's "ego functions" and "defense mechanisms." Social work interventions at that time were usually designated "treatment" rather than "intervention."

As the civil rights movement and the Vietnam war came to the forefront of consciousness in the 1960s, many social workers criticized what they saw as a narrow focus in professional assessments of the time. They called for much greater empha-

Table 5

Multisystem Assessment Incorporating Neuroscience Research
(Complex Adaptive Systems), Late 1990s and Early 2000s

1. What is the client's presenting problem(s)?
   Client's view of the problem.
   View of the problem by significant others.
2. What is the history of the presenting problem?
3. What is the client's current life situation and family history (family, work and school, health issues in family, economic security, religious involvement, social interaction outside family)?
4. How have previous helpers, if any, tried to alleviate the problem(s)? With what results?
5. How do members of the client's family interact with each other?
6. How are stressors from environments outside the family impacting family members and influencing the dynamics between them?
7. How does cultural diversity influence the situation? (Include the worker's subcultures and the organizational subculture of agencies providing services in the analysis.)
8. What environmental and individual risk factors and protective factors are associated with this problem?
9. What neurobiological variables may be involved in the affects, behaviors, and cognitions that seem to be problematic or, conversely, protective, including genetics, brain systems or structures, or neurochemicals?
10. What interventions are known to induce change in visible risk factors and any related, but invisible, internal neurobiological states? What interventions can enhance existing protective factors or create new protective factors?
11. What does the client want? How can we (client and worker) use the information we've compiled in the service of the client's goals and preferences?

---

sis on social conditions (poverty, discrimination, war) and cultural influences, coupled with a shift away from a medical model of *diagnosis* to a systems-oriented assessment. The word *intervention* was now substituted for *treatment*. Many clinicians trained prior to the '60s were not happy with these developments, deploring what they perceived to be the erosion of diagnostic acumen and treatment skills.

With respect to assessment, written evaluations of suspected psychiatric disorders still emphasized the components of the psychosocial diagnosis described above. Psychiatric diagnosis itself continued in the tradition of nonbiologically oriented appraisals of psychosocial factors thought to have caused or contributed to the individual's disorder. Most departments of psychiatry continued in the ego psychology tradition; behavioral clinical practice was the exception rather than the rule. Clinical assessments from both major conceptual frameworks of the day (ego psychology, behaviorism) differed greatly from each other in their understanding of behavior and psychiatric disorders, but shared an emphasis on nonbiological analysis focused on the individual client as an object of therapeutic change.

In social work, general systems theory, although conceptually satisfying to many, was perceived as too divorced from the realities of everyday practice to offer much tangible help. In the 1970s and '80s, the profession enthusiastically embraced a concept that seemed to highlight context and environment while maintaining a focus on individual needs and family dynamics: the *ecological* model of social work practice (Germain, 1982, 1987). Subgroups within the disciplines of psychiatry and psychology (community psychiatry and community psychology) followed parallel paths, emphasizing environmental conditions as well as interpersonal influences as determinants of individual and family behavior (Bagley, 1992; Revenson and Seidman, 2002; Schlosberg, 1976; Smail, 1999; Wilson, Hayes, Greene et al, 2003; Woods, 1975).

During the decade of the 1990s, growing attention to systemic assessment and intervention converged in fields such as epidemiology, community psychiatry, and community psychology. These disciplines simultaneously encompassed a macrosystem perspective and the most micro facet of the person, neurobiological structures, and functions. Their comprehensive, multilevel systems framework reflected the broad view of human functioning formulated decades previously by general systems theorists (Bertalanffy, 1956; see Chapter 5).

The complex adaptive systems model, which had originated in biomedical fields such as cardiology, began to be recognized as an appropriate 21st century paradigm for the fields of psychology and mental health, building on early systems theory but in ways that were more relevant to practice. The complex adaptive systems model was tailor-made for fitting specific neurobiological knowledge into a systems-oriented framework that emphasized the role of environments in shaping mental functions (Gottesman, 2001) (see Table 5). In social work, where many had lagged behind other disciplines in acquiring knowledge about neurobiology and integrating it into a systems framework, the promise of complex adaptive systems won official recognition in a lead article in the *Journal of Social Work Education* (Strohman, 2003; see Chapter 7).

Before we consider ways to carry out assessments from a complex adaptive systems perspective, let's look at the development of social work assessment with respect to mental health issues. To do this, we'll compare questions typically addressed in each of three time phases. Table 4A, p 186, presents questions addressed in both psychosocial diagnosis and systems-oriented assessment during the decades of the 1950s through the 1980s.

Table 4B, p 187, presents the areas where psychosocial diagnosis and systems-oriented assessment may diverge, spanning the '70s, '80s, and early '90s. Table 5, p 189, presents questions that systems-oriented assessments are expected to address today, incorporating neuroscience research.

Here it's important to note that the term *protective factors* refers to the group of variables emphasized in social work's strengths perspective (see Chapter 5). We'll use the terms *strengths* and *protective factors* interchangeably. They refer to individual and environmental characteristics that span the entire range from the most micro to the most macro, influencing individuals, families, and family dynamics. Protective factors can be lifelong or short-term conditions.

When we introduce neurobiology as a new variable interacting with all the other variables considered above, new layers of complexity are added to an already very complex picture. Now we must assess neurobiological conditions as well as psychosocial and environmental conditions. In addition to attending to interacting psychological functions involving continuous exchanges between the individual and social and cultural contexts, we must consider genetics, brain functions, and epigenetic changes through time. How can social workers (or anyone) possibly do anything so complicated? And is there really any need for it?

Several levels of neuroscience knowledge may contribute to our understanding of specific conditions, but fortunately for us, social work assessments require only that we become versed in the topmost layer of this knowledge. Knowledge emanating from deeper layers can enhance our own understanding and expertise, but in most practice situations, it is not critical for competent social work practice.

What does this outermost layer consist of? It is composed of knowledge that we do need pertaining to possible brain structures, systems, and neurotransmitters that mediate the given condition, and knowledge about the effectiveness of interventions that can improve mood, behavior, or thinking as they simultaneously alter the neurochemistry that processes these moods or behaviors.

For example, there are several kinds of depression operating through varying neurobiological structures and processes. We know that a majority of these depressed conditions respond favorably to medications that raise synaptic levels of serotonin (selective serotonin reuptake inhibitors, or SSRIs). Other kinds of depression respond poorly to SSRIs but favorably to monoamine oxidase inhibitors (MAOIs). Yet a third kind of depression requires mood stabilizers and/or anticonvulsants, and can actually be made worse by the use of SSRIs. To help clients suffering from depression, we need information about differences in neurobiological underpinnings of depression and ways of assessing what type of depression a particular client has. Although we don't prescribe, we often have decisive influence on medical personnel who do prescribe but who may hardly know our clients. Their job description often relates mostly to evaluating the need for medication and prescribing drugs likely to be effective for a particular client.

When we compare questions raised in the early 2000s through the lens of complex adaptive systems (see Table 5, p 189) with questions addressed in multisystem assessments prior to the emergence of neurobiology (see Tables 4A and 4B), we find at least three important additional areas of inquiry.

First, what neurobiological variables may be involved in the affects, behaviors, and cognitions that are problematic or, conversely, protective? These variables may include genetics, brain structures (such as the amygdala), or brain chemicals (such as neurotransmitters).

Second, what interventions (such as medication, alternative biological remedies, psychotherapy, support groups, or spiritual practices) may diminish risk factors or enhance protective factors and positively affect visible psychological characteristics and related neurobiological states?

Third, what are the client's preferences with regard to how this information can be used to help?

Let's look at a list of areas of information we seek in a complex adaptive systems assessment. Questions 1-5 pertain to history of the presenting problem, views of the problem by the individual and others, family history, situational factors, family dynamics, and previous experiences with professional helpers. Questions 1-5 are common to the approaches used in all three models (see Tables 4A, 4B, and 5). The order of questions has been modified slightly in Table 5.

Questions 6-8 require the worker to focus on the effects of environmental forces and also to interpret the influence of cultural forces on the individual and the family, including culturally based behaviors and attitudes of the worker and of the agencies providing services. Questions 6-8 are common to multisystem assessments with and without neurobiology, but are seldom used in psychosocial diagnosis.

Questions 9 and 10 require that we look to *recent research* to obtain information that can guide assessment and intervention with particular individuals. Specifically, research information from neuroscience, epidemiology, psychiatry, and social work practice in mental health is needed. Questions 9 and 10 appear only in the multisystem approach that includes neurobiology, i.e. the complex adaptive systems model (see Table 5).

Question 11 requires practitioners to elicit and listen to the client's "voices and choices" (Singh, 2000) as a routine part of assessment. Question 11 is critically important to make proper use of information compiled in the assessment by subjecting it to the guidance of client preferences.

In Chapter 25 we'll look briefly at some of the issues addressed in Chapters 23 and 24 as they play out in a practice situation. A much more detailed practice example (Chapter 27) demonstrates application of a complex adaptive systems approach to contemporary social work practice.

For readers who want more detail and depth, there are many other questions that can advance our own learning about the brain's role in the psychological functions that we experience every day. These experiences occur frequently as we interact with our clients, our own families, and others, and as we think about our own feelings and behaviors. Although answers to these questions may not be required in order to be helpful in any given practice situation, they are important for deepening our thinking and our understanding about human behavior in the social environment.

Examples of these questions follow:

1. How do environmental variables interact through time with neurobiological variables to determine emotions, behavior, and cognition?
2. What interactions appear to be taking place between the

individual's observable behavior or emotions, environmental inputs, and invisible neurobiological events?

3. What evidence is there that neurobiological development and experience are mutually influencing?
4. How does gene expression alter social behavior?
5. How does social experience act on the brain to modify gene expression, brain functions, and even, sometimes, brain structures?

Pennington (2002) has identified four levels of analysis that are required to answer these questions:
a) *brain mechanisms* that alter the developmental process;
b) the *impact of experience on brain development;*
c) specific *neuropsychological processes that are disrupted* by environmental events;
d) *surface behaviors or symptoms* that take place in conjunction with brain mechanisms and brain development.

These levels of analysis apply equally to a presumed "normal" dimension—resilience—and to atypical ("abnormal") psychic functions (Pennington, 2002). Psychiatric diagnosis at the time of writing (2004) is based solely on (d), surface behavior or symptoms referred to as "signs and symptoms" in DSM terminology (APA, 2000).

In a recent review of research that has begun to bridge the gap between neuroscience and clinical practice, Gould and Manji (2004) acquaint us with work in several areas: genetics, epigenetics, profiling gene and protein expression, endophenotypes, neuroimaging, and animal models of psychiatric illness. The term *epigenetics* here refers to regulation of gene activity that is controlled by heritable but potentially reversible changes in gene expression (see Chapter 7). Profiling gene and protein expression refers to describing sets of genes that are transcribed (copied) from DNA into RNA, and sets of genes that are translated from RNA to cells, giving cells instructions about their structures. I refer readers eager for this exciting and challenging information to the authors just cited, and to the many others that can be accessed using databases such as PubMed or PsycInfo.

25
Jenny's Rages
Oppositional 9-Year-Old, Seizures, Bipolar Disorder, and Professional Culture

A 9-year-old girl has severe, intense tantrums at home with her single-parent mother. During these episodes, which are characterized by prolonged rages, she throws things, swings the cat by the tail so that its head bangs against the floor, becomes very physically agitated, and screams at the top of her lungs. The episodes occur at least once every couple of days and last an hour or two. Immediately prior, the child becomes hot and reports feeling light-headed. When the episode passes, the child falls asleep and wakes without any recollection of what happened.

She has been seen in a hospital day treatment facility and an agency day treatment facility, where therapists have diagnosed her with oppositional defiant disorder and have advised the mother that she has poor parenting skills. The mother and I have obtained copies of all medical records and have synthesized them with developmental, medical, and family history to demonstrate that the child qualifies for a diagnosis of bipolar disorder (nowadays sometimes called "mood disorder NOS," because of the diagnostic inadequacies of the current DSM, particularly where children are concerned). Providers, however, are often unwilling to make this diagnosis, regardless of the evidence.

Particularly in complex, difficult cases, each institutional provider wants to affirm the diagnosis assigned by the previous provider, because making a new diagnosis, regardless of the justification, exposes the provider to potentially greater liability in the event of mishaps (there's safety in consensus). This is particularly the case when the pharmacological problems are complex, as with juvenile onset bipolar disorder. The child in question was unstable on medications administered by an outpatient psychiatrist specializing in juvenile onset bipolar. If pharmacology is diagnosis-driven, there's added reason to stick with diagnoses for which there are relatively "safe" medication options.

And the medication question itself, as I always explain to parents, is handled differently depending on the prescriber's philosophy and value system. For example, if mental illness has robbed the child of executive functions, making it impossible for her to perform academically or to play a game of checkers with peers, do you (1) medicate conservatively, because the child is nine and you worry about the potential long-term effects of the neuroleptic, or (2) medicate more aggressively, because the child is suffering not only from the illness's symptoms and the stigma of having the illness, but also from the forgone opportunity to succeed at important developmental tasks of childhood? (The second option presumably entails the hope that medical advances will both improve knowledge of long-term effects of current medications and bring onto the market new, more effective, and more benign pharmaceutical options.)

To sum up, the institutional question is: which gets priority, consideration of the client's needs, or consideration of the institution's exposure to risk? The medication question is: which consideration should govern present assessment, possible future costs or known, here-and-now quality-of-life issues?

Afterthoughts. I don't know whether EEGs are routinely done when children are receiving only day treatment. Certainly they would not be done at an agency; whether hospital day treatment programs order them or not I don't know. And I also don't know whether EEGs are routinely ordered when a diagnosis of juvenile onset bipolar is suspected. In this connection it's important to bear in mind that diagnoses not only *may* but *must* be made even without medical tests. As a social worker, I cannot order tests, but I must supply a diagnosis if my services are to qualify for reimbursement by managed care. Outpatient clinics and agencies under contract to our state's child protective services, which also assign diagnoses, may have medical personnel who ask the primary care provider (outside the hospital or agency) to order one or more tests. However, whether or not the medical personnel are involved with the case, and whether or not tests are ordered, some diagnosis will certainly be assigned. In other words, anyone who wants mental health care from *any* source and is not prepared to pay for it out of pocket must anticipate that the provider *will* assign a diagnosis—and that this diagnosis will enter the patient's medical record somewhere with no indication of supporting evidence.

26
Borderline Personality Disorder[1]
Illustrating Neurobiological Knowledge
for Assessment and Intervention

What is borderline personality disorder (BPD)? After two decades of debate, there's still only limited agreement. Although DSM criteria are widely used (see Part IV, introduction), some scholars of BPD doubt that BPD is a specific disorder (Schmahl, McGlashan, and Bremner, 2002). Instead, they view BPD in terms of dimensions, each with its own neurobiological underpinnings: *affective instability* (fluctuating moods), *impulsivity* (related to such behaviors as substance abuse, binge eating, spending, and aggressive outbursts), *interpersonal stress*, and *dissociation and self-injury*. The first three of these four dimensions also emerged from a validation study of DSM-IV criteria (Blais et al, 1997), suggesting that this dimensional view of BPD might indeed be a meaningful approach to assess-

[1]Grateful acknowledgment is made for permission to include material on practice approaches "Borderline Personality Disorder" in *Adult Psychopathology, Second Edition,* 1999, pp 448-52. Copyright (c) 1999 by Francis J Turner, Editor. Reprinted with permission of *The Free Press*, a Division of Simon & Schuster Adult Publishing Group. All rights reserved.

ment and intervention, a view increasingly endorsed by BPD researchers (Schmahl, McGlashan, and Bremner, 2002). Silk (2000) advocates a dimensional understanding of personality disorders because biological alterations probably underlie *dimensions of behavior* rather than any specific *disorder*.

However, with respect to BPD, I'll refer to the commonly used DSM criteria except when otherwise stated, because I want to keep the review as closely connected as possible to realities of everyday practice.

In this chapter, we'll consider characteristics of BPD and some underlying neurobiological structures and functions associated with these characteristics. Then we'll look at the range of interventions evaluated by research on intervention effectiveness, and implications of this knowledge for social work practice.

People who meet criteria for borderline personality disorder in one or more classification systems are well known to social workers. We see them in almost every type of practice setting, with problems such as suicide attempts and gestures, family violence, substance abuse, eating disorders, reckless spending, and other problems of self-control (Johnson, 1999). These diverse issues require a range of multisystem interventions. We often feel challenged by these clients due to characteristics such as intense hostile-dependent feelings toward the practitioner, overidealization of the social worker alternating with rageful disappointment, suicidal or otherwise violent behaviors, impulsivity, recklessness, and the tendency to terminate treatment abruptly and prematurely when painful issues arise (Stone, 1994; Sadock and Sadock, 2003).

Marsha Linehan, well known as the creator of dialectical behavior therapy, conceives of BPD as five dimensions of a core disorder, psychic dysregulation. These include *emotional dysregulation* (highly reactive emotional responses including episodes of depression, anxiety, irritability, or anger); *interpersonal dysregulation*, characterized by fear of abandonment and chaotic, intense, difficult relationships; *behavioral dysregulation* (extreme impulsivity seen in profligate spending, sex, binge eating, self-mutilation, reckless driving, or suicidal attempts); *cognitive dysregulation* (brief nonpsychotic depersonalization, delusion, or dissociation); and dysregulation of the self (feelings of emptiness or problems with self-identity) (Linehan, 1993).

Intolerance of aloneness, low levels of emotional awareness, inability to coordinate positive and negative feelings, poor accuracy in recognizing facial expressions of emotion, and intense responses to negative emotions are other characteristics of BPD (Levine, Marziali, and Hood, 1997; Gunderson, 1996).

There is agreement that the self-destructive behaviors often engaged in by people with BPD, such as cutting or burning themselves or attempting suicide, are very effective in alleviating psychic pain, especially anxiety and anger (Linehan, 1993). These behaviors are also effective in managing the environment, for example by getting others to admit the person with BPD to a hospital or otherwise to express concern and caring.

Epidemiology. BPD is estimated to occur in 1-2% of the population (Sadock and Sadock, 2003; Torgersen, Kringlen, and Cramer, 2001). Close relatives of persons with BPD have a higher prevalence than average of major depression and substance-use disorders (Sadock and Sadock, 2003). However, studies differ in estimates of the proportion of first-degree relatives of people with BPD who themselves meet criteria for BPD, ranging from negligible to as high as 23% (Dahl, 1994; Torgersen et al, 2001). The estimate that females qualify for the diagnosis twice as often as males is questionable, since males with BPD are often in jail and remain undiagnosed (Cowdry, 1997). As of the mid-1990s, persons with BPD constituted 10-25% of all inpatient psychiatric admissions (Springer and Silk, 1996).

Long-term prognosis appears to be good for the majority of persons diagnosed with BPD. It's a chronic disorder into middle age, by which time the majority of persons formerly diagnosed with BPD no longer meet the criteria. In several studies, mean Global Assessment Scale scores were in the normal range by middle age, and most former patients were working and had a social life.

What treatment could account for this good outcome? Actually, it's unclear how often improvement is due to treatment at all; it may occur as the natural course of BPD. To the extent that treatment during crisis periods can deter suicide, however, it would be essential to buy time by "holding" BPD sufferers until the risk of suicide has passed. About 10% of persons with BPD commit suicide by the 15th year of posttreatment follow-up (Paris, 1993).

The borders of borderline personality. Symptoms and defining characteristics of BPD overlap with symptoms of several DSM-IV-TR diagnoses including other personality disorders, major depression and dysthymia, the spectrum of bipolar disorders, substance abuse, eating, and anxiety disorders, and disorders of dissociation and depersonalization (APA, 2000).

Unstable mood is a defining characteristic of BPD, and depression is a major component of many borderline states. A strong positive family history of *serious affective disorder* is often found among persons meeting criteria for BPD. Many have symptoms characteristic of bipolar, cyclothymic, and unipolar affective disorders. It's sometimes difficult to distinguish ultrarapid-cycling forms of bipolar disorder, in which people are morose, are irritable, and have mood swings, from BPD (Akiskal, 1996). *Substance abuse* also occurs commonly with BPD.

Several studies highlight overlaps between *dissociative* phenomena and BPD (Schmahl, McGlashan, and Bremner, 2002; Zanarini, Ruser, Frankenburg et al, 2000; Kemperman, Russ, and Shearin, 1997). Dissociation is increasingly being recognized as a characteristic of persons with BPD and conversely, among people meeting criteria for dissociative identity disorder (DID), borderline characteristics are very common.

Self-injurious behavior (SIB) is common in BPD and is often associated with relief of intense sadness or other painful states (Schmahl et al, 2002).

Persons with *eating disorders* often also meet criteria for BPD (Palmer, Birchall, Damani et al, 2003). Using blood platelet measures of serotonin and monoamine oxidase activity to assess monoamine functions, Verkes, Pijl, Meinders et al (1996) found that people with bulimia could be subdivided into two groups: those with and those without BPD. Persons with bulimia and BPD resemble suicide attempters with BPD in levels of anger, impulsivity, and biochemical characteristics.

Neurologic dysfunction in persons meeting the criteria for BPD is common (Schmahl et al, 2002). It is important to distinguish the terms *neurologic* and *neurobiogical*. All psychic phenomena have neurobiological underpinnings, whereas the term *neurologic* designates specific subtypes of neurobiological dysfunctions that are diagnosed and treated by the specialty of neurology rather than psychiatry. These distinctions are becoming increasingly fuzzy as knowledge about the neurobiology of psychiatric disorders advances, but they are rooted in medical tradition.

Etiology. Research indicates that no matter which diagnostic criteria are used, borderline personality is the outcome of a heterogeneous developmental course involving interacting biological and environmental risk factors. The components appear to be biological vulnerability (predisposition for affective instability, impulsivity, or other borderline characteristics) interacting with a childhood history of lack of validation, loss, abuse, or other unidentified environmental inputs (Cowdry, 1997; Linehan, 1993).

Borderline dimensions, neurobiological functions, medications that address these functions. Evidence is accruing rapidly that specific neurochemical abnormalities underlie atypical personality characteristics in BPD (Markovitz, 2004). Given the high degree of overlap between characteristics of BPD and so many other disorders, it shouldn't be surprising that there is wide variation in brain chemistry and functions among people who meet criteria for the BPD diagnosis.

It is well documented that high percentages of people who meet criteria for BPD now take psychotropic medications (Zanarini, 2004). Since particular neurochemistries are associated with specific psychological characteristics, pharmacological treatments of BPD now usually select specific behavioral anomalies to target medication choices. Since my earlier reviews of BPD, views about the role of medication in BPD have changed from support by only a minority of providers, to widespread endorsement of medication as a major intervention for BPD (Johnson 1988; Markovitz, 2004).

Although effective medications are often discovered by trial and error, advances in knowledge about underlying neurobiological functions and structures that mediate dimensions of BPD are helping providers use medications more accurately. Studies now show that genetics make a major contribution to BPD outcomes, as genes interact with environmental inputs throughout years of development. Torgersen et al (2001) found an overall heritability of 0.69 for BPD in a study of 92 monozygotic and 129 dizygotic twin pairs. High levels of genetic influence were found for affective instability, impulsivity, and self-injurious behavior (ibid). Affective instability and self-injurious behavior had previously been identified as having a strong genetic basis by Livesley, Jang, Jackson et al (1993), whose research also showed genetic underpinning of identity problems.

Prominent dimensions of BPD have been studied with respect to brain regions, structures, systems, and neurotransmitters. *Affective instability* is typically understood as extreme sensitivity to emotionally meaningful environmental events, rapid shifts in emotional state (mood swings), emotional intensity, and difficulty returning to emotional equilibrium (Linehan, 1993; APA, 2000). People with BPD often differ from people with major affective disorders in that their mood swings typically involve becoming irritable quickly in response to *external* factors. For persons with BPD who also fall in the spectrum of bipolar disorders, however, irritability and rapid mood swings may be a response to internal neurobiological states inherent in the illness.

Although there is now considerable knowledge about neurobiological bases of the two groups of affects seen in affective instability (the lows in depression, the highs in mania), the underlying functions that make these affects *unstable* is not yet clear.

The down phase of affective instability appears as major depression or, if less severe, dysthymia (chronic depressed mood). Low levels of serotonin are clearly associated with depression, as are low levels of norepinephrine (noradrenaline). Neurons that discharge serotonin have presynaptic receptors for *norepinephrine* that appear to regulate the amount of serotonin released.

Dopamine may also play a role in depression. Drugs that reduce dopamine concentration such as reserpine and diseases that reduce dopamine concentration such as Parkinson's are associated with depressive symptoms, whereas drugs that increase dopamine concentration such as tyrosine, amphetamine, and bupropion reduce depressive symptoms (Sadock and Sadock, 2003). The depressive characteristics of anhedonia (inability to feel pleasure) and loss of motivation appear to be related to deficiency of dopamine in the mesolimbic dopamine pathway, a brain system involved strongly in transmission of the experience of pleasure (Klimek, Schenck, Han et al, 2002).

The medial prefrontal cortex (MPFC), one of the brain structures that regulates emotions and stress, has shown alterations in depression, one of the conditions that frequently overlap with BPD. Neuroendocrine functions have been clearly related to depression through structures such as the hypothalamus, which has a central role in regulating endocrine functions. Other transmitters have also been implicated in the physiology of depression, such as gamma-aminobutyric acid (GABA), vasopressin, endogenous opioids, and glutamate (Sadock and Sadock, 2003).

Brain tissue of suicide victims has shown reduced levels of serotonin as well as an increased number of serotonin receptors in the prefrontal cortex and the hippocampus, which is believed to be an attempt by the organism to adapt to chronic low levels of serotonin (Stanley, Molcho, Stanley et al, 2000).

The up phase in affective instability is suggested by evidence pertaining to the manic phase of bipolar disorder (shown as hyperactivity, elated mood, excitement, grandiosity, or irritability). The up phase appears to involve serotonin systems, serotonin transporters in presynaptic function, and possibly the benzodiazepine binding site on the gamma-aminobutyric acid receptor in bipolar disorder (see p 78), as well as other neural structures too detailed to have been included in this volume. Readers wanting further information about the neurophysiology of depression and mania can search PubMed.

Perhaps the best evidence for specific neurobiological events in affective instability comes from the responses of people with BPD to medications. *Antiepileptic* drugs *valproate, lamotrigine*, and *topiramate* have been helpful for persons with BPD, consistent with the view that some forms of BPD are bipolar spectrum disorders (Markovitz, 2004). These medications are front-line medications for bipolar disorder. Valproate was effective in a placebo-controlled study for most symptoms in persons meeting criteria for both BPD and bipolar II, except for symptoms of depression (Frankenburg and Zanarini, 2002). Valproate decreased irritability and impulsive aggressive behavior in clients who had failed to respond to other antiaggressive agents such as selective serotonin reuptake inhibitors in an open trial (Kavoussi and Coccaro, 1998). It was also helpful for various symptoms in a small-sample study of persons with BPD without psychotic, depressive, or bipolar disorder (Hollander, Allen, Lopez et al, 2001). Lamotrigine was notable for positive results with three clients who had very severe BPD and a history of failures on other medications (Pinto and Akiskal, 1998; Rizvi, 2002).

Impulsivity, another core feature of BPD, is a deficit in the inhibition of response. The serotonin system and the medial prefrontal cortex play major roles in impulsivity (Schmahl et al, 2002; Markovitz, 2004; Paris, Zweig-Frank, Kin et al, 2004). An inverse relationship between impulsive aggression and neurochemical indices of serotonin system function, and between a life history of aggressive behavior and metabolites of serotonin, have been reported in several studies.

Impulsivity is a major constitutional predisposition in BPD, regardless of whether there is a history of trauma (Schmahl et al, 2002). Various forms of impulsive behavior have been associated with platelet monoamine oxidase activity (Brown, Goodwin, Ballenger et al, 1979; Linnoila, Virkkunen, Scheinen et al, 1983). Linnoila and colleagues showed that serotonin levels were reduced only if aggression is impulsive, rather than premeditated. When hyperactivity and impulsivity act together in an individual with BPD, self-destructive behavior often comes into play.

The medial prefrontal cortex regulates impulse control and inhibition of responses to external stimuli (Schmahl et al, 2002). Using positron emission tomography (PET scan), De La Fuente, Goldman, Stanus et al (1997) found relatively low levels of glucose consumption in the premotor and frontal cortices, the anterior cingulate cortex, and thalamic, caudate, and lenticular nuclei in persons with BPD, indicating significant underactivity in these regions.

Selective serotonin reuptake inhibitors (SSRIs) have been recommended for treatment of depression, impulsivity, and aggression in persons with borderline personality disorder, based on findings from 10 open studies and 3 double-blind placebo-controlled studies (Markovitz, 2004; Zanarini, 2004). Because placebo-controlled studies of SSRIs have proven inconclusive due to the large response to placebo, a larger-scale study was needed to distinguish medication effects from placebo effects. Rinne, van den Brink, Wouters et al (2002) carried out the largest and most comprehensive double-blind, placebo-controlled, randomized study to date of the effects of an SSRI for the treatment of moderately severe BPD in 38 women using the SSRI fluvoxamine. The authors measured rapid mood shift, impulsivity, and aggression. The medication produced a strong and long-lasting reduction in rapid mood shifts that were not helped by placebo, but had no more effect than placebo on impulsivity and aggression.

However, it has also been demonstrated that impulsivity and aggression can be attenuated with SSRI treatment. Researchers have succeeded in reducing aggression and normalizing prefrontal cortex metabolism in 10 persons with borderline personality disorder using fluoxetine (New, Buchsbaum, Hazlett et al, 2004). PET scans verified that improvements as measured on the Overt Aggression Scale were accompanied by significant increases in metabolic rates in two specific regions of the prefrontal cortex (Brodmann areas 11 and 12). It is not known whether the difference in response to fluvoxamine and fluoxetine with respect to aggression in these studies is due to differences between the two drugs or other factors.

For individuals with frequent psychotic symptoms, *atypical antipsychotics* have shown the following effects. Olanzapine benefited BPD clients with self-injurious behavior and other BPD symptoms except for depression, which was not alleviated. Clozapine recipients showed decreases in self-injury and hostility and improved social adaptability. Risperidone reduced impulsive-aggressive behavior and was associated with a 13-point improvement on the Global Assessment of Functioning (GAF) score. Quetiapine was helpful for self-injurious behavior. These antipsychotic medications were clearly useful for some BPD symptoms, but not effective for treating depression in persons with BPD (see Markovitz, 2004, and Zanarini, 2004, for more detailed analysis of these studies). Clients typically take a combination of medications, since no single drug is effective for every type of symptom. Weight gain is often a problem with these antipsychotics.

Another core dimension of BPD, *interpersonal stress,* refers either to a protracted developmental history of stress (such as chronic neglect, invalidation by important adults, or repeated abuse), or to a single intense episode or period (such as war combat or a violent sexual assault). Single episodes or periods occur more frequently in cases of posttraumatic stress disorder, whereas protracted developmental history is more characteristic of persons with BPD.

In persons with BPD, the strength of an individual's hyper-reactivity to the environment, which often manifests itself as hypersensitivity in interpersonal situations, appears to be mediated through noradrenergic mechanisms (Figueroa, Silk, Huth et al, 1997).

Neuroimaging studies have even shown a few specific reactions to particular interpersonal experiences. For example, in people with BPD there were significant increases in blood flow in parts of the prefrontal cortex and decreases in the anterior cingulate in response to scripts depicting situations of abandonment. A study of rhesus monkeys after separation from their mothers showed similar activation in the *same region of the prefrontal cortex* as had been seen in people with BPD reacting to the abandonment scene (Rilling, Winslow, O'Brien et al, 2001). These studies show a clear-cut similarity in emotional processing of real or imagined abandonment by humans and rhesus monkeys in a very specific brain region.

Dissociation is a frequently endorsed symptom in people with both PTSD and BPD. Dissociative phenomena include *depersonalization* (people feel detached from their own senses and from surrounding events, as if they were outside observers), derealization (sense of unreality), and *analgesia* (lack of feeling) (Phillips and Sierra, 2003). These dissociative experiences can arise from underlying neurobiological abnormalities, both inborn and those induced by trauma, or effects of drug including phenylcyclidine (PCP), ketamine, and hallucinogens such as lysergic acid dyethylamide (LSD) and mescaline (Schmahl et al, 2002).

The thalamus seems to play a major role in the processes that lead to dissociative states (Schmahl et al, 2002). It acts as a sensory gate or filter that modulates access of sensory information to the cortex, amygdala, and hippocampus. Interestingly, several of the relay stations of somatosensory pathways have high concentrations of opioid receptors. This suggests that *endogenous opioids* (opioid neurotransmitters occurring naturally) play a role in modulating sensory information processes, thus generating dissociative states. The hippocampus has been implicated as well through a correlation between dissociation and hippocampal atrophy (ibid). Naltrexone, an opioid antagonist, also reduced dissociative symptoms, relating to the involvement of endogenous opioid systems in dissociation (Bohus, Landwehrmeyer, Stiglmayr et al, 1999).

Cowdry (1997) has noted the similarity of dissociative states to complex partial seizures, suggesting a therapeutic role for anticonvulsants. Factor analysis has suggested a connection between dissociation and self-injurious behavior. Some observers believe that the reason self-injury often does not induce pain is because it stimulates excessive activity of endogenous opioids.

The serotonin system is also thought to be involved in dissociative phenomena as well as self-injurious behavior. Medication responses bear this out. Fluoxetine has reduced depersonalization and self-injurious behavior, as has sertraline (see for example Schmahl et al, 2002, p 81).

Medications that have been helpful with various aspects of BPD include an *opioid antagonist, modafinil,* and *omega-3 fatty acid*. Naltrexone, an opioid antagonist used successfully with some individuals addicted to alcohol or opiates, led to a cessation of self-injurious behavior for 32 weeks in an open trial (Griengl, Sendera, and Dantendorf, 2001). Modafinil, used to treat people with narcolepsy (falling asleep suddenly in the middle of the day), was given to 12 clinically depressed people with BPD who experienced lethargy. A third of the group showed marked improvement, a third moderate improvement, and a third little to no improvement in sleepiness (Markovitz and Wagner, 2003). Omega-3 fatty acid was given to women in a double-blind placebo controlled trial and is thought to be helpful to some individuals with aggression and impulsivity, but statistical improvement by the women receiving the supplement did not differ from improvement with placebo. Both groups showed substantial declines in aggression and depression (Zanarini and Frankenburg, 2003).

Several neurotransmitters and neuromodulators (brain chemicals that modify neurotransmission without meeting all the criteria for neurotransmitters) are involved in regulation of eating behavior in animals and have been implicated in symptoms such as depression and anxiety in humans with *eating disorders* (Mauri, Rudelli, Somaschini et al, 1996).

Antidepressants have been effective in reducing frequency of binge eating, purging, and depressive symptoms in persons with bulimia, even in cases of chronic persistent bulimia in patients who had repeatedly failed courses of alternative therapies, and even in persons with bulimia who do not appear depressed (ibid).

Central serotonergic receptor-blocking compounds such as cyproheptadine, an antihistamine, can cause marked increase in appetite and body weight. For anorexia nervosa, zinc supplementation may be helpful in combination with other approaches.

Other studies pertaining to neurobiological underpinnings of BPD have tested relationships between BPD and hormonal responses (Steinberg et al, 1997); BPD and sleep disorders (Akiskal, Judd, Gillin et al, 1997); association between platelet monoamine oxidase activity and stable personality traits such as impulsiveness, monotony avoidance, and aggressiveness (Stalenheim et al, 1997); and a possible neural basis for the phenomenon of "splitting" so often observed in persons with BPD (Muller, 1992). Splitting refers to seeing some people as all good and others as all bad, and idealizing the former while vilifying the latter.

In general, Cowdry (1997) observes that psychotropic medications in treatment of BDP occasionally result in marked improvement, but more often serve to modulate affect enough to make daily experiences somewhat less disruptive. In most cases, medication is only one weapon in an arsenal of individual and group interventions to help persons with BPD develop coping skills.

What can we make of this complicated, confusing information? Schmahl and colleagues conclude that the complex symptoms of BPD are best understood as a combination of alterations in different neurobiological systems (2002, p 81). Each person with BPD must be assessed to identify the unique combination of features that may be responsive to different kinds of interventions for this particular individual.

In multiple regressions of potential predictor variables for the borderline diagnosis, *interpersonal sensitivity* was the only significant predictor of the diagnosis (Figueroa, Silk, Huth et al, 1997). These authors believe that at least in some cases, interpersonal sensitivity may be the inborn constitutional condition with which environmental forces (such as traumatic events) interact to lead to BPD. That is, biologically based interpersonal sensitivity makes such individuals specially vulnerable to trauma. Empirical evidence leads to the conclusion that BPD is an outcome arrived at by interactions between a vulnerable (hyperbolic) temperament, a traumatic childhood (broadly defined), and/or a triggering event or series of events (Zanarini and Frankenburg, 2003). In a particular situation, any one of these three categories may contribute most strongly.

Delayed responses to severe psychological trauma (seen in symptoms of PTSD) present paradoxical symptoms: continuing elevation of epinephrine and norepinephrine response mediating anger and fear, together often with *normal* levels of hypothalamic-pituitary-adrenal (HPA) axis activity (Henry, 1997). A normal level of HPA axis activity is counterintuitive; we would expect it to be elevated in response to trauma (see Chapter 19).

Reexperiencing the trauma and the arousal may be associated with dysfunction of the locus coeruleus, amygdala, and hippocampal systems. In addition, dissociation of the connections between the right and left hemispheres may account for the alexithymia (lack of awareness of one's emotions or moods) and failure of the cortisol response that so often follows severe psychological trauma (see Chapter 19) (Henry, 1997). In this condition, it appears that the right hemisphere no longer fully contributes to integrated cerebral function. Children with damage to the right hemisphere lose critical social skills, adults lose a sense of relatedness and familiarity. Henry postulates that these losses of social sensibilities may account for the lack of empathy and difficulties with bonding found in antisocial personality disorder (ASPD) as well as in BPD (ibid).

Zanarini reviewed medications in use as of the early 21st century (2004). She found that most of the medications studied in double-blind placebo-controlled trials were efficacious. She disputes the view that polypharmacy recommended by some BPD scholars is necessary, given that some individual drugs positively influence multiple troublesome characteristics of BPD. In addition, she believes choice of medications should be guided by tolerability and safety more than by symptom targeting. There is disagreement about this issue among leading BPD researchers, but all agree that medications are important and effective for BPD (Markovitz, 2004). Most were useful to treat symptoms of the core dimensions of BPD, affective dysregulation and impulsive aggression.

Environmental contributors to BPD characteristics. Linehan (1993) identifies a specific kind of environment, which she calls "invalidating," that interacts with biological factors (genetic, intrauterine, or physical trauma at an early age, especially head trauma) to give rise to BPD. Invalidating environments convey the message that our responses to life situations are invalid, inappropriate, or incorrect. In such environments, all problems are viewed as motivational in origin—"If only you were to try harder, the problem would be solved." Typically such environments tell children to "calm down." However, it is very difficult for such children to calm down. These environments fail to recognize the biologically based emotional vulnerability and overly sensitive responses that may lie behind children's inhibitions, irritability, or frequent crying in infancy.

Paris (1997) disputes the belief that adult psychopathology arises primarily from environmental factors, because personality traits are heritable, children are often resilient to the long-term effects of trauma, all studies of trauma in personality-disordered persons suffer from retrospective designs, and only a minority of patients with severe personality disorders report severe childhood trauma.

According to Paris, the effects of trauma in the personality disorders can therefore be better understood in the context of gene-environment interactions (i.e. epigenetic factors). Interaction is key, because many biologically vulnerable persons do not develop BPD, and many persons with histories of childhood trauma or loss do not develop BPD. When a child has a biological vulnerability, it might be possible to prevent a later onset of BPD by recognizing the vulnerability early, supporting parents, and educating them about the child's special, idiosyncratic needs. These vulnerable children require parenting efforts different from and/or greater than the ordinary parenting behaviors that succeed with typical children. At present, however, few if any preventive assessments or interventive programs are in place.

Role of the family. Until the 1980s, parents of persons with BPD were assumed by professionals to have caused their children's disorders. In an early review, Gunderson and Englund (1981) found little evidence to support the literature's messages that parents are culpable. They suggested that the tendency of persons with BPD and their therapists to see the mothers as "bad" may be a function of the splitting mechanisms so frequently used by persons with BPD. They might, in fact, be projecting nonexistent negative behaviors or affects onto their mothers in the process of reconstructing their pasts.

However, many practitioners were oblivious to the evidence and continued to conduct psychotherapy based on an assumption of parental culpability. A few other scholars in the early '80s, however, concurred with Gunderson and Englund. Palombo (1983) observed that persons with BPD experience the world as hostile and chaotic and believe that their caretakers have failed to provide a benign environment.

> Regardless of whether or not the parents did fail, [they] were in all likelihood helpless to provide such an environment for the child. . . . [The] most benign,

responsive, caring parents may in that sense be utter failures from the perspective of the child because his needs were not adequately attended to. Conversely, neglectful parents raising a competent, well-endowed child might be experienced as loving and responsible.

[The person with BPD] could, and often does, pin the blame for his suffering on those around him. From his perspective, they are the causes of it. Since [the parents] are perceived as powerful and mighty, they cannot be absolved of the blame for permitting his suffering to occur. Myths are then created about the terrible things that parents did to the child, even though in reality the parents may have struggled mightily to provide for the child (Palombo, 1983, pp 335-336).

In their study, Gunderson and Englund (1981) had expected to find empirical support for the beliefs prevailing in the late 1970s that family psychopathology for persons with BPD was characterized by overinvolved, separation-resistant, dependency-generating mothers. Instead, they found that parental underinvolvement was much more common. Neglectful parenting was also implicated in a recent study (Zanarini et al, 1997). In comparison with people with other personality disorders, three family environmental factors were significant predictors of a borderline diagnosis: sexual abuse by a male noncaregiver, emotional denial by a male caregiver, and inconsistent treatment by a female caregiver. The results suggested that sexual abuse was neither a necessary nor a sufficient condition for the development of BPD and that other childhood experiences, particularly neglect by caretakers of both genders, represent significant risk factors. However, since the majority of children who experience any of these risk factors do not develop BPD, other variables clearly must be present to account for the later development of BPD.

Gunderson, Berkowitz, and Ruiz-Sancho (1997) caution us

against one-sided assessment of BPD.

> Much of the preceding literature about the families of borderline patients derived solely from reports provided by the borderline patients and rarely included the families' perspectives. When you consider that borderline patients are by nature often devaluative and that they often find on entering treatment that devaluing past caretakers—past treaters as well as families—is a way to ignite the ambitions and enthusiasm of the new candidates for becoming their caretakers, it is disturbing in retrospect that we have not been more suspicious about their accounts of their families (Gunderson et al, 1997, p 451).

Two decades ago psychiatrist Kenneth Terkelsen (1983) apologized publicly to parents of persons with schizophrenia for his unfair and denigrating views about parents' supposed roles in causing their children's illnesses. Fourteen years later, Gunderson published a similar apology with respect to parents of people with BPD.

> I...was a contributor to the literature that led to the unfair vilification of the families and the largely unfortunate efforts at either excluding or inappropriately involving them in treatment. So it is with embarrassment that I now find myself presenting a treatment [psychoeducation] that begins with the expectation that the families of borderline individuals are important allies of the treaters (Gunderson et al, 1997, p 451).

What psychosocial interventions are effective? As of the early 21st century, studies of treatment effectiveness emphasize medication; variants of cognitive-behavioral therapy, notably dialectical behavior therapy (DBT); and the value of "holding" actions by therapists. Holding refers to worker behaviors communicating that he/she is a stable, reliable person who cares about the client, is consistently supportive, and weathers the emotional and behavioral storms of the person with BPD (Waldinger, 1987; Buie and Adler, 1982). In addition to client-related behaviors, social workers must engage in systems-oriented work to ensure that clients are best served by the mental health and other service systems. The practice situation described in Chapter 25 illustrates the need for systems-oriented action by social workers.

With respect to individually targeted interventions, emphasis has shifted from insight-oriented reflective therapies, widely used during the 1980s, to skills training targeted on the most common difficulties of persons with BPD, such as intense anger, self-destructive acts, excessive drug use, binge eating and purging, or reckless spending (Springer and Silk, 1996). Treatment components (inpatient or outpatient) may involve some combination of individual and group psychoeducation, behavioral strategies and skills training (such as anger management, relapse prevention, social skills training), medication, peer support groups, family psychoeducation and support, cognitive therapy, individual psychodynamic counseling, environmental changes, and advocacy. BPD is a chronic condition and requires a wide range of intervention options in a model emphasizing ongoing availability and intermittent active intervention.

In addition, specific problem behaviors of persons with BPD arising from mood or impulse control dysfunctions are targets of treatment. Psychoeducation and cognitive-behavioral approaches are first-line interventions (Linehan, 1993; Gunderson et al, 1997). Insight-oriented psychodynamic psychotherapy appears to have been effective for a relatively small subgroup of persons with BPD who are generally likable by others, motivated, psychologically minded, focused, free of overwhelming impulsivity and substance craving, and without a history of "grotesquely destructive" early environments (Stone, 1987). However, impulsivity and/or cravings are among the defining characteristics of BPD, requiring structured, behaviorally oriented interventions to help persons with BPD control impulses that threaten their ability to maintain jobs or intimate relationships.

Psychoeducation for persons with BPD and their family members is essential, as information is empowering (Gunderson et al, 1997). The question "Why am I this way?" or "Why is my family member this way?" is a salient preoccupation even if not articulated. Palombo (1983) referred to the BPD sufferer's "maddening sense of inadequacy at coping with an imperfectly understood environment" (p 331).

To help people with BPD understand their own internal reactions as well as the effects of inputs from the environment, workers can give information about the role of biological factors in specific aspects of the person's borderline illness and about the ways in which the environment may interact with biological vulnerabilities to cause symptoms and suffering.

For example, when the person with BPD has a history of ADHD, head injury, or other neurological dysfunction, it is helpful to say "You know, your ADHD [head injury, epilepsy] seems to have the effect of making you fly off the handle easily" ["get distracted," "say things you're sorry for," "look for excitement all the time"]. When the client expresses sadness or hopelessness, the worker can point out that the client seems to have a tendency toward depression that interacts with stressful life events.

It is often useful to suggest readings that can help persons with BPD understand their own vulnerabilities and explain treatment options. Broad-based psychoeducation usually includes information about the disorder; benefits and risks of medications, side effects, and signs that the medication requires adjustment (larger or smaller doses, or a different medication); alternative treatment options; and information about financial assistance, health benefits, community resources, employment assistance, support groups, and respite care.

Skills Training and Relapse Prevention: Dialectical Behavior Therapy (DBT).

Dialectical behavior therapy was originally developed to treat women with BPD who are chronically parasuicidal (engage in self-mutilation such as cutting their skin, but do not wish to die). DBT has been adopted in several state departments of mental health as the treatment of choice for BPD. A decade after the first randomized controlled trial of DBT, it remains the only outpatient psychotherapy with demonstrated efficacy for women with BPD (Robins and Chapman, 2004).

DBT has shown success in reducing self-destructive behaviors in people with BPD, reducing hospital admissions, and improving social adjustment (Robins and Chapman, 2004). Linehan (1993) explains that standard cognitive behavior therapy, like psychodynamic therapy, has a change orientation: it strives to change behavior through learning and experience. Where these therapies fail, according to Linehan, is in a lack of acceptance of clients as they are. Treatment based on the goal of change alone may reinforce clients' inability to accept themselves unless an acceptance-as-you-are component is added.

DBT begins with a posture of total acceptance (called *radical acceptance*) borrowed from Eastern religions such as Zen. The word *dialectical* refers to the resolution of polarity and tension between opposites. Persons with BPD must accept themselves as they are and simultaneously try to change. Intervention is geared toward a balance involving supportive acceptance combined with change strategies. The treatment is behavioral because it focuses on skills training, collaborative problem solving, contingency clarification and management, and the observable present, and is directive.

Individual visits with a therapist and weekly group sessions follow a manualized set of steps. First, the therapist and client must reach a collaborative agreement focusing on suicidal or other self-harming behaviors. Will the person with BPD agree to stop doing these things? Before other issues can be addressed, the individual must be willing to agree to stop hurting or threatening to hurt herself or himself. The worker/therapist emphasizes that there are only two options: to accept one's condition and stay miserable, or to try to change it.

The social worker engages with the client in active problem solving, responds with warmth and flexibility, and validates the patient's emotional and cognitive responses. For example, the worker recognizes that the client's coping behaviors (such as cutting, burning, or attempting suicide) have been very effective or functional for that person by supplying predictable relief from pain. Treatment focuses on figuring out ways to avoid or escape from pain other than hurting oneself.

The worker frequently expresses sympathy for the client's intense pain and sense of desperation, creating a validating environment while conveying a matter-of-fact attitude about current and previous self-destructive behaviors (Linehan, 1993). The worker reframes suicidal and other dysfunctional behaviors as part of the client's learned problem-solving repertoire and tries to focus the counseling process on positive problem solving. In individual and group sessions, workers teach emotional regulation, interpersonal skills, tolerance of emotional pain, and self-management skills. They openly reinforce desired behaviors to promote progress by maintaining contingencies that shape adaptive behaviors and extinguish self-destructive behaviors. They set clear limits to their availability for help. They emphasize building and maintaining a positive, interpersonal, collaborative relationship using social work roles of teacher, consultant, and cheerleader.

DBT has been adapted for several other populations. It has shown effectiveness or promise with substance use disorders, suicidal adolescents, eating disorders, inmates in correctional settings, depressed elderly persons, and adults with attention-deficit hyperactivity disorder (Robins and Chapman, 2004). In a recent comparison of DBT and a non-behavioral "treatment by experts" (TBE), well-known practitioners in DBT were nominated by community mental health leaders to do the comparison group treatment. Both groups showed significant improvement overall with comparable rates of suicide. However, women in the DBT group had a significantly lower frequency of serious suicide attempts and use of emergency room and inpatient services, and experienced less than half the rate of treatment dropouts from the initially assigned therapists of women in the TBE group (Linehan, Comtois, Brown et al, 2002).

Because their psychic pain is extreme, people with BPD may engage repeatedly in self-destructive behavior as they learn that it reduces emotional pain. It is not yet known how this mechanism works. Studies have compared persons with BPD who experience physical pain during self-injury with those who do not. Those reporting no physical pain show lower electroencephalogram activity and more mechanisms of dissociation during self-injurious behavior than those who report feeling pain. Even during self-reported states of calmness, persons with BPD show a lower perception of physical pain than healthy controls. The extent to which not feeling pain might be related to impaired cognitive ability to distinguish pain is not known (Bohus, Limberger, Ebner et al, 2000; Russ, Campbell, Kakuma et al, 1999; Kemperman, Russ, and Shearin, 1997).

DBT as developed by Linehan and her followers would seem to offer aspects of the holding environment in addition to its other features. Advocates of the holding environment believe that healing occurs not through interpretation, but by being a stable, consistent, caring, nonpunitive person who survives the client's rage and destructive impulses and continues to serve this holding function (Waldinger, 1987, p 270). According to this view, experiential factors are more important than psychological interpretations. Buie and Adler (1982) advocated implementing the holding function through such supports as hospitalization when needed, extra appointments, phone calls between sessions, provision of vacation addresses, and even postcards sent to the client while on vacation. Linehan, however, recommends setting firm limits on the worker's availability, but emphasizes a worker posture of reliability, stability, and consistent support.

Persons with BPD typically move from crisis to crisis, so following a behavioral treatment plan may be difficult, especially if this plan involves teaching skills that are not obviously related to the current crisis and that do not promise immediate relief. Therefore, behaviorally oriented workers have developed psychoeducational group treatment modules using group process to teach specific behavioral, cognitive, and emotional skills (Linehan, 1993).

Group therapy as part of a multimodal treatment package has been rated as highly effective by inpatients with BPD (Leszcz, Yalom, and Norden, 1986; Linehan, Armstrong, Suarez et al 1991). Psychodynamically oriented group approaches, however, have not demonstrated effectiveness (Springer and Silk, 1996).

Psychiatric rehabilitation models address the long-term and chronic nature of BPD by providing a validating environment, the opportunity to enhance skills, and a mechanism to accept input from families. Specific skills are taught, such as learning to accurately label one's own emotions, general problem-solving skills, and skills to enhance self-esteem and deal with anger, as spelled out in Linehan's DBT (Links, 1993). Hospitalization in acute psychiatric settings is viewed as not very desirable, as it may be too emotionally stimulating and not validating. Other settings must be found, such as a community agency or a work environment, that mesh with the patient's characteristics and that can create a sense of validation.

Day treatment or partial hospitalization can sometimes bridge the gap between hospitalization and outpatient treatment for people with BPD who need more intensive care than outpatient individual and group therapy. Miller (1995) describes four characteristics of successful programming in a day setting for persons with BPD: affect facilitation, holding without overcontaining, ensuring client safety, and providing focused time-limited treatment (three weeks in Miller's program).

However, all settings short of round-the-clock supervised hospitalization entail considerable risk of suicide. Suicide also occurs even on locked units, but less frequently than in part-time community care. On locked units, patients at high risk for suicide are monitored with 15 minute observations around the clock.

To address the suicide risks, DBT therapists emphasize taking a strong position that they cannot save clients from themselves; only the individuals can ensure their own safety (Linehan, 1993; Miller, 1995). Some practitioners and scholars of BPD argue that hospitals take away responsibility for individual self-care in persons with BPD by removing "sharps" and other implements of suicide, a view that fits with current managed care practices of quick discharge to out-of-hospital care. As with any treatment, there are undoubtedly clients for whom this policy works, others who may be casualties through suicide who conceivably might not have died in long-term hospital treatment.

Work with families of persons with BPD. Psychoeducation and family support groups have emerged as interventions of choice in work with families of persons with BPD (Gunderson et al, 1997). Recent research on BPD suggests that average or "good enough" mothering may be insufficient to protect a child with neurobiological vulnerabilities from developing borderline characteristics. Conversely, resilient children at high risk due to traumatic environments may grow into well-functioning, successful adults. A borderline outcome may be the cumulative result of interactive biological vulnerabilities and life stresses, with or without parental inadequacies.

The blame/shame stigma now beginning to dissipate for other major psychiatric disorders, such as schizophrenia and bipolar disorder, has been slower to diminish for BPD. This is probably because persons with BPD seldom have signs of psychosis but frequently exhibit irritating or upsetting behaviors. In the absence of evidence to the contrary, parents are often assumed to be at fault when their children grow up with problematic characteristics.

Parents of persons with BPD need to be educated that BPD is a neurobiological disorder, not retribution for toxic parenting. They need the support of others in similar situations to break down their sense of stigma, self-blame, and hopelessness, and they need training in ways to cope with behaviors typical of

BPD. As Gunderson has pointed out, parents should be viewed as collaborators in the treatment process, not objects of therapy (Gunderson et al, 1997).

Implications for Social Work Practice

How does this new expanded knowledge fit with our roles as case managers, therapists, family consultants, patient and family advocates, and social and political activists? All the contemporary approaches mentioned above fall within the purview of social work practice and fit with a complex adaptive systems approach. Even with respect to psychotropic medication, social workers perform all critical functions except for writing prescriptions, doing medical evaluations, and interpreting laboratory reports.

Prior to intervening, it is imperative that we familiarize ourselves with current research-based knowledge about characteristics and etiologies of BPD, what interventions are effective for what types of BPD clients, and under what conditions they are effective. Research published during the past few years has given us tools for understanding and treating BPD in ways that were formerly beyond our capability. However, without valid knowledge about the diverse forms and different underlying characteristics of BPD, and without training in skills targeted at troublesome or dangerous characteristics of the disorder, there is little reason to expect that our treatments will be effective.

Research suggests that if we can prevent suicide in the crisis phase of the disorder, individuals have a good chance at a satisfying life later. However, the recovery process is usually very slow and gradual. We should have a reasonable expectation that the illness will be chronic for an extended time and that relapses are to be expected. Meanwhile, until the risk of suicide is diminished, intensive therapeutic efforts are needed. At the same time, as social workers we can remind ourselves and our clients with BPD that tools of self-destruction are always available in the community, that we cannot save them, and that ultimately only they can choose to stay safe.

The natural course of BPD, gradual diminution of symptoms over a period of years, underscores the need for ongoing structure and monitoring in the community and for short-term protected environments (hospitals or some other kind of safe haven) where clients can stay until the immediate threat of suicide has passed. Expansion, rather than the contraction of such services that is now happening under the aegis of managed care, must be vigorously advocated by social workers in the political sphere as well as on an individual basis. At the time of writing, the American public has begun to voice frustration and anger about the impact of current cost-cutting measures throughout managed-care-dominated health care systems. Some social workers are taking an active role in this effort, but we need to do much more.

Skills necessary to work with persons with BPD and their families are teachable and learnable. However, it appears that training in these skills is not available in some social work programs, requiring workers to get it elsewhere. In our view, generic social work skills simply are not adequate to address the specific challenges of work related to this painful and often dangerous condition. Given the prevalence of this disorder across social work settings, social work educators should think of ways to bring therapeutic skills training modules directly into graduate and undergraduate curricula. Agencies do in-service trainings in specialized therapies such as DBT. A solution to the lack of behavioral skills-based training in some social work programs might be to offer electives taught by agency trainers well versed in DBT or other interventions.

In conclusion, I summarize specific skills that workers need.

1. The ability to provide psychoeducation for persons with BPD and their families in group and individual venues, conveying up-to-date information about characteristics and etiology of the disorder, helping clients explore the meaning of this information for their own lives, giving information about treatment options and community resources, and explaining potential benefits and risks associated with these.

2. The ability to create a validating environment, so as to help

the person with BPD accept himself or herself while at the same time working actively to change.

3. The ability to clearly specify and enforce the conditions under which client and practitioner will work together, especially the boundaries of treatment.

4. Cognitive-behavioral skills for teaching impulse control, anger management, relapse prevention, and self-soothing, which can be conveyed well in groups as well as individual sessions.

5. With respect to medication, practitioners must access up-to-date information so they can share this information with clients and their families. Together with clients and family members, they can effectively monitor the client's response to a medication and report to the prescribing physician when necessary. Workers do not need to memorize drug information, as it changes continually, but they should become adept at doing computer searches on PubMed to obtain state-of-the-art information about drug effectiveness and side effects. Because these searches can be done on the Internet with only a minimal expenditure of time, technology has made keeping up with current knowledge infinitely easier than it was only a few years ago. The easiest method is to get on-line access to databases such as PubMed or PsycInfo at home, or if not feasible, in college or public libraries.

6. Finally, skills in advocacy and political action are required for social workers to join with other activist groups fighting to make services available to thousands of service consumers and their families when they need them. At the time of writing, the gap between these goals for service delivery and the reality of their availability is great. All of the above should be available to persons with BPD on an as-needed basis, not only for crisis management but also to provide the continuity of a relationship with a consistent, caring person who can weather the client's outbursts of hostility and self-destructive threats and acts, and can convey hope that a happier future is possible.

These skills are the same as those I advocated earlier (Johnson, 1999, pp 451-452), except for one aspect. A major change is that advances in electronic databases have made information about neurobiological aspects and treatment effectiveness much easier to access. Learning about state-of-the art interventions to help our own clients is now a few clicks of the mouse away, once we learn the relatively simple skills of accessing research abstracts on line. Guidelines for evaluating the validity of research studies are offered by Gibbs (2003). The implication of this change is that workers can now go straight to the best experts around the world to gain assistance in assessment using current information, and help with intervention using research on what interventions work for what types of clients, under what conditions. Although much remains to be done in filling out this template for practice knowledge, we've made amazing strides, and progress continues.

27
Pat, Food Addiction, and Thoughts of Death

The client and family ecosystems evaluation that follows was written by a clinical social worker. The evaluation follows the guidelines for multisystem assessment as presented in Table 5, Chapter 24, incorporating neuroscience research. At the end of the evaluation, the social worker uses the data compiled in the history to present his assessment and explain what interventions he thinks may be helpful. His agency does not use this protocol. The attending physician's brief discharge note is included. The social worker wrote the detailed report for this chapter, disguising the material to avoid identifying the clients and himself.

Discharge Note May 27th, 2004
Pat W, a 27-year-old 98-pound Caucasian female was admitted May 22nd, 2004, in acute crisis due to dehydration and elec-

trolyte imbalance. She has a 12-year history of bulimia nervosa and intermittent anorexic episodes. This is her third hospitalization in a crisis condition over the 12-year period. She was hydrated intravenously and treated with cyproheptadine to stimulate weight gain. Vital signs are now stable. She is being discharged to outpatient care pending verification of insurance eligibility.

Evaluation and Recommendations
Pat W, May 27th, 2004
Bristol Community Mental Health Center

1. Presenting problem(s); views of the problem by client and significant others. Pat is a 27-year-old Caucasian woman who has suffered from eating disorders (bulimia and intermittent anorexia) since the age of 15. She has recently been admitted to the inpatient psychiatric unit at the Bristol Community Mental Health Center, a state-funded multiservice facility for psychiatric patients in a small industrial town in a midwestern state. This is her third hospitalization over the 12-year period since the inception of disordered eating behaviors. Pat's 5-day length-of-stay allowed by her insurance for persons with her diagnosis has elapsed, and she is about to be discharged. The treatment staff is not in agreement about recommendations for ongoing treatment.

Some members of the treatment staff view Pat's problem as stemming from failed attachment in childhood and ambivalent relationships with both parents. However, previous psychotherapy geared to working through these issues has failed. Pat views her problem as inability to control insatiable urges to binge and purge. She is very cognizant that this problem is ruining her life, but feels helpless to do anything about it. She expresses the wish to be dead, but denies having formulated any specific plans for committing suicide.

Nevertheless, all three of her hospitalizations have occurred in response to life-threatening medical crises brought on by starving herself or intensive bingeing and vomiting. The fact that she has allowed herself to get into a dangerous physical condition on repeated occasions suggests at least passively suicidal behavior.

Pat has revealed other important facets of her view of the problem. She reports often not being able to feel the way other people do, sometimes even feeling detached from her body. "I see other people having fun and enjoying things and laughing. I just can't seem to feel anything." She has complained through the years of feeling "empty." Pat does not accept professionals' views that she and her mother did not attach or bond properly, or that her conflicts with her father during adolescence were a major contributor to her emotional fragility. She sees her feelings of emptiness and sadness as something arising within herself that she cannot explain.

2. History of the presenting problem(s).

Pat was overweight most of her childhood and into adolescence. Some of her classmates referred to her as "Fatty Patty." At age 15, she was rejected by a boy she'd had a crush on for two years, and learned via the grapevine that he had told friends that he had no interest in Pat because she was fat. This experience was excruciating for her. She responded by intensive dieting, losing 60 pounds over the summer months by eating only "rabbit food" and jogging several miles a day in the hot sun.

Her schoolmates responded approvingly to her new persona when school reopened in the fall. She was now slender, had dates for the first time, but still struggled with food. On Christmas day, her mother, according to Pat, prepared all her favorite foods, and Pat stuffed herself. "My mom's cooking is awesome!" She looked at herself sideways in the mirror, saw a very distended belly, and believed this condition was permanent. All her suffering to lose weight, blown in one huge binge. It was at that moment it occurred to her that she could get rid of all the food by vomiting. She induced vomiting, and immediately felt an overwhelming sense of relief. She said she had found a "magic bullet." She began to eat voraciously again. She vomited more and more frequently.

In the ensuing 12 years, she meticulously counted calories and tried to eat no more than 900 per day, but also continued to binge and purge. She attended college, worked in department store sales, and had an active social life, but described herself as still being in a prison of deprivation, bingeing, and purging. Sometimes Ella, Pat's mother, would bake a bag of cookies and bring it with a gallon of milk to Pat at her apartment. Pat would eat the entire amount in one sitting, then lie to her mother about who had eaten it.

Pat went to the extent of vomiting calorie-free soda, even though she knew this behavior was irrational since diet sodas would not make her fat. She usually hid her eating behaviors from her family, friends, her fiancé during a year-long engagement, and eventually therapists whom she saw intermittently starting her sophomore year in college.

3. Client's current life situation and family history: family, work and school, health issues in family, economic security, religious involvement, social interaction outside family.

Pat is the oldest of four, three girls and a boy, all of whom were overweight during their growing-up years. Pat did very well in school until the teen years, attended college for two years while working part time, then left to work full time. At the time of her current hospital admission, she was working as manager of the women's clothing division of a department store, but had lost considerable work time just prior to the hospitalization. Pat and her brother, Freddie, are very close. Freddie has taken an active role in trying to connect Pat with sources of help, and in supporting their parents through years of grief and stress with Pat's condition.

Pat lives in the town next to her parents' town. She is single, currently without a partner, but was engaged for about a year when she was 25 to Evan, who was nurturing, devoted to her, and willing to extend himself to help her overcome her eating compulsions. He himself had overcome obesity in high school and was very sympathetic to her fear of regaining lost weight.

To her family and friends, Evan seemed like a wonderful match for Pat. However, she broke their engagement. It appeared that Evan could not fill the emptiness that she was experiencing.

Pat's mother Ella was one of 12 children in a poor family, her father Edwin one of five. Maternal and paternal grandparents worked in a local factory and eked out a living for their large families. Ella and Ed told the social worker they've been very happy together for 29 years, but have suffered greatly because of Pat's illness. Ella becomes tearful when recounting how she and Ed lie awake nights wondering what's happening with Pat. They brought many pictures of happy times when the children were young: Ed with Pat on his lap reading a picture book, Ed helping Pat learn to ride a bicycle, Ella baking cookies with Pat and her sister Joanne as small children, and family gatherings.

Ed continues to work in a small local factory, and Ella works part time in a drugstore. Ella and Ed attend a local Catholic church, as do their other two daughters. Pat, however, is no longer religious, but attends church to be with the family on major holidays. The parents live in a small house in the middle of their town, where they have been most of their married life. Ella is delighted with the comfort they are able to afford, which her family of origin could not: central heating and an indoor toilet.

The family socializes mostly with relatives, many of whom live nearby. Both Ella and Ed conveyed to the social worker a sense of family support, cohesion, and affection.

Apart from Ella's high blood pressure and Pat's moods and eating disorder, family members are in good health. Ella has been known to forego her blood pressure medicine at times when Pat was living at home and eating enormous amounts of food, because there wasn't enough money for both. When questioned about this, Ella said to the social worker, "Wouldn't any mother give up something for one of her children?"

4. Previous experiences with help.
Pat first came to the attention of mental health professionals

while in college. Her brother Freddie had suspected Pat was vomiting for some time, confronted her, but until Pat was hospitalized in medical crisis (life-threatening electrolyte imbalance), she had been unwilling to seek professional help. Her mother, in a family interview with a social worker, said she was shocked when she found out about Pat's bingeing and purging at the time of the first hospitalization. "I don't know why I never thought of that, it just didn't occur to me." Pat acknowledged in a moment of honesty that she is a talented actress and manages to fool almost everyone, most of the time, about her bingeing and purging.

Records from previous hospitalizations did not specify what treatments were given beyond individual psychotherapy, milieu therapy on the unit, daily group therapy, and medication to promote hydration and appetite. After the longest hospitalization, which lasted several weeks, Pat appeared to her family to have given up the binge-purge behavior, but reverted some months later. Pat has been in and out of outpatient treatment over the past 12 years, but typically discontinues when her therapist becomes confrontative about her bingeing and purging, which she attempts to deny and conceal.

The Bristol Center's policy is to treat persons diagnosed with schizophrenia or bipolar disorder with medication, but to use pharmacotherapy only rarely with nonpsychotic illnesses. Insight-oriented therapy is emphasized for "internalizing" disorders (involving subjective internal states such as depression and anxiety). Behavior modification is the clinic's frontline intervention for "externalizing" disorders (involving disruptive behaviors typical of attention-deficit hyperactivity and intermittent explosive disorders). During this current hospitalization, Pat has been hydrated, fed intravenously, and treated with cyproheptadine to stimulate appetite, but no ongoing medication has been prescribed.

5. Family interactions and dynamics.
Food has always been a central part of the family's life. Ella is

of short stature but has always been very overweight. "I grew up the hard way, we all worked very hard, and you had to get nourishment to keep up your strength." Ed grinned as he recounted how he fell in love with Ella on their first date, in which she cooked him a meal of roast beef, yorkshire pudding, and fudge cake. Ella once decided to try to lose weight but did not follow through. Pat has helped her mother learn to read and write, and they appear to be very fond of each other. Ella says the other three children are "easygoing like me," enjoy life, and take things as they come.

Ed and Pat, however, are alike in that they are serious, perfectionistic, and easily upset. Ed reports that at age three, Pat already had to do everything perfectly. Her parents said she was not convinced when they tried to explain that she doesn't have to be "the best." During her adolescence, Pat and her father had repeated conflicts over her social life. "I'm from the old school, but she just thought she could come and go as she pleased, and she wore those tight sweaters that showed everything." Ed believes that teenagers should do what their parents ask "because I say so." Pat responded to her father's injunctions about curfews, clothes, and other social issues by flouncing out the door on dates with boys whom Ed did not like. Pat had outbursts of rage, sometimes breaking dishes. These outbursts were often a response to what she perceived as irrational restrictions, but Pat also had outbursts of anger in relation to situations at school, and later, in the workplace. Her episodes of intense anger, then, have been fairly pervasive rather than being limited to her relationship with her father. Ed and Ella say they've done a lot of soul-searching over the years. They wonder whether they might have caused Pat's illness.

6. Environmental stressors.
Pat does not identify major sources of environmental stress in her life except when she becomes too ill to work, causing anxiety about income. However, she has never been without money for long, because she has been very successful in department store work, is highly valued by her employer, and is not in the

line of fire in times of retrenchment. For her parents, Pat's illness is the major source of stress, pain, and grief. Although they are of modest means, they feel fortunate to be so much better off than their families were as they were growing up. Ella takes her high blood pressure in stride, considering it a normal part of growing older. Although having to choose between buying food and medicine would be considered a major stress by most people, Ella does not seem too troubled about it. Her major distress is Pat's unhappiness. Both Pat and her parents have what they consider to be acceptable health insurance through their employers, except for medication. However, recent cutbacks are causing them some worry about how they will manage in the event one of them becomes very ill.

7. Cultural influences on the client and family members; worker's subculture; organizational subculture of agencies providing services.

Cultural influences on the family.

Societal promulgation of thinness, through mass media and social institutions, as a requirement for attractiveness.

Importance of food in this family.

Belief (Ella's) that providing abundant food is an aspect of loving parenthood.

Rural poverty background.

Roman Catholic religion.

Generational differences in attitudes toward sex, role of parents in adolescent decision-making.

Worker and agency subcultures. The agency subculture, i.e. its beliefs about etiology and treatment, were summarized above. My own views as Pat's social worker are somewhat different from those of my agency, because I was trained in a teaching hospital where research was used to inform assessment and intervention. That's seldom the case at Bristol. My colleagues go to workshops and retreats led by big names in the field, but usually these are pop psychology gurus. I know I sound like a snob, but sometimes I get upset that staff make decisions about clients' lives by quoting some well known therapist ("Dr S says. . . .") without ever going to see what current research says. How do they know that this particular client's problems are comparable to the situations the gurus presented in their workshops? Where I was trained, you had to defend your treatment recommendations by referring to research.

8. Environmental and individual risk factors and protective factors.

Risk factors

Pat's temperament: tense, anxious, volatile, easily angered, perfectionistic, obsessive.

Ella's belief that eating a lot and being fat is "OK," as it was for her in her youth when survival needs took precedence over vanity.

Ella's wish to express love to Pat by giving her large amounts of delicious home-cooked food (a risk factor in this situation, but would be a protective factor in another family).

Media images of slender glamorous women, mixed with intensive corporate advertising to promote sale of junk foods.

Cultural differences between Ella's and Ed's generation and the high-tech media culture that has shaped their children's values and beliefs.

Worker and agency subcultures.

Based on what I explained under (7) above, I guess I'd have to say that the agency's lack of attention to research is a risk factor.

Protective factors

Pat's intelligence.

Pat's competence at her job.

Ella's and Ed's love for Pat.

Ella's and Ed's devotion to each other and support for each other.

Freddie's concern and help for Pat and their parents.

Stable long-term residence in same place.
Health of Ed, Pat's siblings.

Potential protective factors
Practice interventions yet to be tried:
    Educating parents, Pat, siblings.
    Trials of medication for Pat.
    Exploring possible state or federal government medical assistance.
    Psychotherapy for Pat, specifically DBT.
    Support group for family members.

Some of these protective factors will only happen if I succeed in connecting Pat with an outpatient facility that will give her a trial of meds and DBT, neither of which has been tried yet.

Social Work Notes
Based on the data compiled above, my impression is that Pat meets criteria for DSM-IV-TR diagnoses of bulimia, intermittent anorexia, dysthymia, intermittent major depression, and probably borderline personality disorder (impulsivity in one major area rather than two, affective instability, chronic feelings of emptiness, intense anger, as well as possible transient dissociation or identity disturbance). She also meets Linehan's (1993) criteria for borderline personality disorder, in that she has 4 of 5 types of dysregulation (emotional, behavioral, cognitive, and dysregulation of the self) (see Chapter 26). In addition, she has obsessive-compulsive characteristics (obsessive thinking about food, compulsive eating and purging behavior) and is in an addictive state (out-of-control frequent overconsumption of food, inability to stop this behavior despite severe adverse consequences in her life, attempting to conceal the behavior and/or deny that it is taking place).

To find information for Questions 9 and 10 on the Multisystem Assessment protocol, I did searches of current research on these diagnoses relating to neurobiological underpinnings and effectiveness of interventions. Methods for conducting searches were drawn from Leonard Gibbs, Evidence-Based Practice for the Helping Professions (see Chapter 8 in this volume).

I listed sources about eating disorders that were not included elsewhere in this book. In addition, I drew on my own experience to recommend systems-oriented interventions that might help, because there doesn't seem to be much research targeted on work with external systems beyond the family.

I spent much more time doing these searches in Pat's behalf than is usually necessary with my clients, because Pat has a long history of chronic illness that does not seem to have responded well to treatment, she has symptoms of several different DSM-IV-TR disorders, and she's obviously suffering terribly. Her illness is really complicated and challenging. I want to try to find ways to help her break out of this seemingly endless cycle of bingeing, purging, and despair. The abstracts that I obtained in my searches gave me some ideas about interventions that might help Pat, despite her 12-year history of treatment failures.

9. Neurobiological variables possibly involved in Pat's emotions, behaviors, and cognitions, including genetics, brain structures, and brain chemicals.

I began by considering temperaments, Pat's and those of members of her family. In this regard, Pat and her father Ed are similar to each other in that they are serious and intense. Pat is inflexible, and Ed may resemble Pat in this regard. Pat is unlike either parent in that she is quick to anger and extremely perfectionistic. Ella and Ed agree that Ella, Freddie, and the other two girls are very different: easygoing, laid back, and usually full of fun. As Ella says, "the others take after me."

Then I pulled up abstracts from PubMed that pertained to Pat's psychological characteristics. I searched under depressed mood, suicide, eating disorder, impulsivity, obsessive-compulsive, feelings of emptiness, and inability to feel. I learned that impulsivity of the kind that's related to eating compulsions is characterized by "impaired response inhibition," which means inability to control urges. Brain imaging has shown that it's related to abnormalities in the frontal cortex and cingulate cortex (smaller volume and reduction of gray matter) in persons with borderline personality as compared with typical people. Depressed mood is related to serotonin functions and also, in some people, to norepinephrine functions, also called noradrenaline. Low levels of serotonin are associated with suicidal ideation and even seem to predict suicide attempts.

Eating disorder abstracts pointed to serotonin dysfunction in anorexia and bulimia, reduced blood flow in the temporal lobes and cingulate cortex, loss of brain gray and white matter in anorexia which is partially reversible with treatment, and overlap with obsessive-compulsive traits[1]. Obsessive-compulsive characteristics are also related to inadequate levels of serotonin, specifically in the basal ganglia. I discovered that functional imaging studies have shown similar metabolic abnormalities in the prefrontal cortex and caudate nucleus in both anorexia nervosa and OCD, suggesting that the underlying neurobiology is similar in some ways[2]. I also learned that serotonin status in bulimia may correspond to impulsivity (low serotonin activity) and obsessive thinking about food[3,4].

One study found that anorexia involves an underlying neuronal circuit abnormality[5]. Another study showed that people with either anorexia or bulimia responded to food stimuli in a similar way in the same brain region (the ventromedial prefrontal cortex), which did not happen in healthy women, suggesting that there was a common underlying disease process in these two disorders[6]. The authors of that study noted that similar responses to inappropriate stimuli take place not only in eating disorders, but also in obsessive-compulsive and addictive disorders, which may account for similar features of behavior in all these conditions.

Genetics are involved too. Researchers found an association between anorexia and 3 genes, all of which code for serotonin receptors[7]. What was really interesting for Pat's case was that in people with bulimic symptoms, a variant of a serotonin transporter gene coincided with significant elevations in affective instability, behavioral impulsivity, and greater likelihood of meeting criteria for borderline personality disorder[8]. All of this sounds like Pat. Could it be that Pat has this genetic abnormality? Probably we'll never know. But suppose there were a treatment that could offset or block the action of that gene? Might she be released from her 12-year prison sentence?

I couldn't find any information using keywords "feelings of emptiness," except that it is associated with depression. I did find an abstract on inability to feel, using the keywords "emotional numbness." Pat's description of her inability to feel like other people and her sense of being apart from the rest of the world sounded like depersonalization to me, which is defined as an alteration in the perception or experience of the self[9].

10. Interventions known to induce change in visible psychological characteristics and/or underlying neurobiological states.

My preliminary assessment. I disagree with the Bristol Center's not using psychotropic medication in Pat's case. I also disagree with the view that optimal treatment for Pat is to help her gain insight into her longings for intimacy and attention, and her feelings of emptiness at not having these. From Pat's descriptions of her previous therapies, it sounds as though that has already been tried, once for at least two years. Pat doesn't think it helped her at all, although she says she tried to respond to her therapist's efforts. This exploration involves looking back into her childhood at presumed painful lacks in the attachment bond between herself and her mother. I concur with Pat's own view that she and her mother are strongly bonded. Pat's problems are undoubtedly exacerbated by her mother's plying her with food, and that's an area where her family needs education, but I don't believe the cause is attachment failure in childhood and adolescence. Rather, they stem mostly from Pat herself, probably developed over time through interactions between her temperament, the roles of food in her family, and media-driven culture.

I pulled up abstracts on effectiveness of treatments that addressed several of Pat's characteristics: binge eating, vomiting, obsessive thinking about food and impulsive eating, depression, and feelings of depersonalization. All of these characteristics except depersonalization seemed to be associated with inadequate levels of the neurotransmitter serotonin. SSRIs (selective serotonin reuptake inhibitors) have been successful in treating depression, obsessive-compulsive symptoms, and bulimia, even when the person with bulimia isn't depressed.

It seems to me to make sense for her to have a solid trial with an SSRI, maybe fluoxetine because it has been shown to be effective in so many studies, but obviously I don't know about the minutiae of the different meds. That would be for the prescribing physician to decide. *My job is to raise the questions.* Maybe if the doc doesn't seem familiar with some of the recent research I could give him the folder of abstracts, without being too offensive. I'd say something like "I know how busy you are, so perhaps you'd like to take a look at these abstracts I just pulled up."

Binge eating has also been helped by two other classes of drugs, anticonvulsants (especially topiramate) and a serotonin and norepinephrine reuptake inhibitor (sibutramine). Atypical antipsychotics (especially olanzapine) have been found to be effective in some cases as well. I've added these sources to my list (see Sources 10-16 below).

With respect to psychotherapy, studies of the outcome of dialectical behavior therapy strongly suggest it could help Pat. I believe she should engage in it on an outpatient basis over an extended period of time, maybe a year. Other psychotherapies don't seem to have worked on any long-term basis. Pat has not had DBT, and DBT has the support in the abstracts for the view that a combination of medication and DBT is likely to be more effective than either one alone.

I was very interested to find that a full DBT program was evaluated in a pilot study of 7 women with both BPD and eating disorder[16]. "Full" includes weekly individual therapy, weekly skills training group, out-of-hours telephone contact by experienced clinicians who receive the approved intensive training program in DBT, and a weekly consultation group for the therapists. The authors point out that one of the strengths of DBT is the clear supportive framework that it supplies for patients and clinicians alike. In this pilot program, unlike many treatment programs for eating disorders, there were no dropouts. All patients seemed to benefit, and most were neither eating disordered nor self-harming at follow-up. This was a small pilot study and clearly larger, more systematic evaluations are needed, but the program showed considerable promise for helping a population known to be very difficult to treat.

Finally, Pat's family needs support and education about the illness, like all families with loved ones who have a psychiatric illness. I'd like them to attend a family support group for family members of persons with eating disorders or borderline personality disorder, whichever one we manage to find in this area. I might even have to start such a group if there is none.

11. Client preferences. I've explored Pat's preferences for treatment following her discharge. She feels desperate and hopeless, and is willing to try anything that she hasn't already tried that might help. I'm in a difficult position, because I want to connect her with what I think can help her the most, but my views are at odds with those of my agency administrators. I've been looking into outpatient treatments that are offered within any kind of reasonable distance from where Pat lives. The closest facility is 60 miles away, but it uses medication for a wide range of conditions and it also has a small DBT program. My view is that it would be well worth Pat's time to do that driving, which she'd probably have to do twice a week, once for group and once for individual therapy. It looks as though her insurance will cover it. She'll have to decide.

Sources for Multisystem Evaluation of an Eating Disorder

1. Brewerton TD (2004). 9th Annual Meeting of the Eating Disorders Research Society. Expert Opinion on Investigational Drugs 13(1):73-78.

2. Murphy R, Nutzinger DO, Paul T, and Leplow B (2004). Conditional-associative learning in eating disorders: a comparison with OCD. Journal of Clinical and Experimental Neuropsychology 26(2):190-199.
3. Steiger H, Israel M, Gauvin L, Ng Ying Kin NM, and Young SN (2003). Implications of compulsive and impulsive traits for serotonin status in women with bulimia nervosa. Psychiatry Research 120(3):219-29.
4. Serpell L, Livingstone A, Neiderman M, and Lask B (2002). Anorexia nervosa: obsessive-compulsive disorder, obsessive-compulsive person ality disorder. Clinical Psychooogy Review 22(5):647-69.
5. Lask B (2003). The neurobiology of early onset eating disorders. Abstract of the 9th Annual Meeting of the Eating Disorders Research Society, Ravello, Italy, October 1-4.
6. Uher R, Campbell J, and Treasure J (2003). Functional neural correlates of symptom provocation in eating disorders. Abstract of the 9th Annual Meeting of the Eating Disorders Research Society, Ravello, Italy, October 1-4.
7. Bergen AW, Welch R, Yeager M et al (2003). Indoleamine candidate gene association analysis in the price foundation anorexia-nervosa affected-relative pair dataset. Abstracts of the 19th Annual Meeting of the Eating Disorders Research Society, Ravello, Italy, October 1-4.
8. Steiger H, Joober R, and Israel M (2003). Behavioral and affective dys-regulation, serotonin activity, and serotonin transporter gene polymor-phism in bulimic and nnormal-eater women. Abstract of the 9th Annual Meeting of the Eating Disorders Research Society, Ravello, Italy, October 1-4.
9. Phillips MI and Sierra M (2003). Depersonalization disorder: a function al neuroanatomical perspective. Stress 6(3):157-165.
10. Appolinario JC and McElroy SL (2004). Pharmacological approaches in the treatment of binge eating disorder. Current Drug Targets 5(3):301-07.
11. Walsh BT, Fairburn CG, Mickley D, Sysko R, and Parides MK (2004). Treatment of bulimia nervosa in a primary care setting. American Journal of Psychiatry 161(3):556-61.
12. Tammela LI, Rissanen A, Kuikka JT, Karhunen LJ, Bergstrom KA, Repo-Tiihonen Naukkarinen H, Vanninen E, Tiihonen J, and Uusitupa M (2003). Treatment improves serotonin transporter binding and reduces binge eating. Psychopharmacology (Berl) 170(1):89-93.
13. McElroy SL, Arnold LM, Shapira NA, Keck PE Jr, Rosenthal NR, Karim MR, Kamin M, and Hudson JI (2003). Topiramate in the treat ment of binge eating disorder associated with obesity: a randomized placebo-controlled trial. American Journal of Psychiatry 160(2):255-61.
14. Kotler LA, Devlin MJ, Davies M, and Walsh BT (2003). An open trial of fluoxetine for adolescents with bulimia nervosa. Journal of Child and Adolescent Psychopharmacology 13(3):329-35.
15. Pederson KJ, Roerig JO, and Mitchell JE (2003). Towards the pharma-

cotherapy of eating disorders. Expert Opinion in Pharmacotherapy 4(10):1659-78.
16. Palmer RL, Birchall H, Damani S, Gatward N, McGrain L, and Parker L (2003). A dialectical behavior therapy program for people with an eating disorder and borderline personality disorder—description and outcome. International Journal of Eating Disorders 33:281-286.

References Part IV

Akiskal HS (1996). The prevalent clinical spectrum of bipolar disorders: beyond DSM-IV. Journal of Clinical Psychopharmacology 16(2 Suppl 1):4S-14S.

Akiskal HS, Judd LL, Gillin JC, and Lemmi H (1997). Subthreshold depressions: clinical and polysomnographic validation of dysthymic residual and masked forms. Journal of Affective Disorders 45(1-2):53-63.

American Psychiatric Association (2000). Diagnostic and Statistical Manual of Mental Disorders, 4th ed, text revision. Washington, DC: American Psychiatric Association.

Bagley C (1992). The urban environment and child pedestrian and bicycle injuries: interaction of ecological and personality characteristics. Journal of Community and Applied Social Psychology 2(4):281-289.

Bertalanffy L von (1956). General systems theory. General Systems 1:3.

Blais MA, Hilsenroth MJ, and Castlebury FD (1997). Content validity of the DSM-IV borderline and narcissistic personality disorder criteria sets. Comprehensive Psychiatry 38(1):31-37.

Bohus MJ, Landwehrmeyer GB, Stiglmayr CE, Limberger MF, Boehme R, and Schmahl CG (1999). Naltrexone in the treatment of dissociative symptoms in patients with borderline personality disorder: an open-label trial. Journal of Clinical Psychiatry 60:598-603.

Bohus MJ, Limberger MF, Ebner U, Glocker FX, Schwarz B, Wernz M, and Lieb K (2000). Pain perception during self-reported distress and calmness in patients with borderline personality disorder and self-mutilating behavior. Psychiatry Research 95(3):251-60.

Brown GL, Goodwin FK, Ballenger JC, Goyer PF, and Major LF (1979). Aggression in human correlates with cerebrospinal fluid amine metabolites. Psychiatry Research 1:131-39.

Buie D and Adler G (1982). The definitive treatment of the borderline patient. International Journal of Psychoanalysis and Psychotherapy 9:51-87.

Cowdry R (1997). Borderline personality disorder. NAMI Advocate Jan/Feb, 8-9.

Dahl AA (1994). Heredity in personality disorders—an overview. Clinical

Genetics 46(1 Spec No):138-43.

Dean B (2004). The neurobiology of bipolar disorder: findings using human postmortem central nervous system tissue. Australia and New Zealand Journal of Psychiatry 38(3):135.

De La Fuente JM, Goldman S, Stanus E, Vizuete C, Morlan I, Bobes J, Mendlewicz J (1997). Brain glucose metabolism in borderline person ality disorder. J Psychiatr Res. 1997 Sep-Oct 31(5):531-41.

De Silva P, Menzies RG, and Shafran R (2003). Spontaneous decay of compulsive urges: the case of covert compulsions. Behavior Research and Therapy 41(2):129-37.

Figueroa EF, Silk KR, Huth A, and Lohr NE (1997). History of childhood sexual abuse and general psychopathology. Comprehensive Psychiatry 38(1):23-30.

Frankenburg FR and Zanarini MC (2002). Divalproex sodiium treatment of women with borderline personality disorder and biopolar II disorder: a double-blind placebo-controlled pilot study. Journal of Clinical Psychiatry 63(5):442-6.

Germain CB (1987). Human development in contemporary environments. Social Service-Review 61(4):565-80.

Germain CB (1982). Teaching primary prevention in social work: an eco- logical perspective. Journal of Education for Social Work 18(1): 20-28.

Gibbs L (2003). Evidence-Based Practice for the Helping Professions: A Practical Guide with Integrated Multimedia. Pacific Grove, CA: Thomson Brooks/Cole.

Gottesman II (2001). Psychopathology through a life span-genetic prism. American Psychologist 11:867-78.

Gould TD and Manji HK (2004). The molecular medicine revolution and psychiatry: bridging the gap between basic neuroscience research and clinical psychiatry. Journal of Clinical Psychiatry 65(5):598-604.

Griengl H, Sendera A, and Dantendorf K (2001). Naltrexone as a treatment of self-injurious behavior—a case report. Acta Psychiatrica Scandinavica 103:234-236.

Gunderson JG (1996). The borderline patient's intolerance of aloneness: insecure attachments and therapist availability. American Journal of Psychiatry 153(6):752-58.

Gunderson JG, Berkowitz C, and Ruiz-Sancho A (1997). Families of bor- derline patients: A psychoeducational approach. Bulletin of the Menninger Clinic 61(4):446-57.

Gunderson JG and Englund DW (1981). Characterizing the families of borderlines. Psychiatric Clinics of North America 4(1):159-168.

Henry JP (1997). Psychological and physiological responses to stress: the right hemisphere and the hypothalamo-pituitary-adrenal axis, an inquiry into problems of human bonding Acta Physiol Scand Suppl. 640:10-25.

Hollander E, Allen A, Lopez RP, Bienstock CA, Grossman R, Siever LJ, Merkatz L and Stein DJ (2001). A preliminary double-blind, placebo-controlled trial of divalproex sodium in borderline personality disorder. Journal of Clinical Psychiatry 62(3):199-203.

Johnson HC (1999). Borderline personality disorder. In FJ Turner, ed, Adult Psychopathology, 2nd ed, pp 430-56. New York: Free Press.

Johnson HC (1988). Where is the border? Current issues in the diagnosis and treatment of the borderline. Clinical Social Work Journal 16(3):243-260.

Kavoussi RJ and Coccaro EF (1998). Divalproex sodium for impulsive aggressive behavior in patients with personality disorder. Journal of Clinical Psychiatry 59(12):676-80.

Kazdin A (2001). Behavior Modification in Applied Settings, 6th ed. Belmont, CA: Wadsworth Thomson.

Kemperman I, Russ MJ, and Shearin E (1997). Self-injurious behavior and mood regulation in borderline patients. Journal of Personality Disorders 11(2):146-157.

Klimek V, Schenck JE, Han H, Stockmeier CA, and Ordway GA (2002). Dopaminergic abnormalities in amygdaloid nuclei in major depression: a postmortem study. Biological Psychiatry 52(7):740-8.

Leszcz M, Yalom ID, and Norden M (1986). The value of inpatient group psychotherapy: Patients' perceptions. International Group Psychotherapy 85: 411-433.

Levine D, Marziali E, and Hood J (1997). Emotion processing in border line personality disorders. Journal of Nervous and Mental Diseases 185(4):240-246.

Linehan MM (1993). Cognitive-behavioral treatment of borderline personality disorder. New York: Guilford.

Linehan MM, Armstrong HE, Suarez A, Allmon D, and Heard H (1991). Cognitive-behavioral treatment of chronically parasuicidal borderline patients. Archives of General Psychiatry 48:1060-1064.

Linehan MM, Comtois KA, Brown MZ, Reynolds SK, Welch SS, Sayrs J, and Korslund KE (2002). DBT vs non-behavioral treatment by experts in the community: Clinical outcomes at one year. In SK Reynolds (Chair), University of Washington treatment study for borderline personality disorder. Symposium conducted at the annual meeting of the Association for Advancement of Behavior Therapy, Reno, NV.

Links PS (1993). Psychiatric rehabilitation model for borderline personality disorder. Canadian Journal of Psychiatry 38 Feb(Suppl. 1):535-538.

Linnoila M, Virkkunen M, Scheinin M, Nuutila A, Rimon R, and Goodwin FO (1983). Low cerebrospinal fluid 5-hydroxyindoleacetic acid concentration differentiates impulsive from nonimpulsive violent behavior. Life Sciences 33:2609-2614.

Livesley WJ (2004). Introduction to the special feature on recent progress in the treatment of personality disorders. Journal of Personality Disorders 18(1):1-2.

Livesley WJ, Jang KL, Jackson DN, and Vernon PA (1993). Genetic and environmental contributions to dimensions of personality disorder. American Journal of Psychiatry 150(12):1826-31.

Markovitz PJ (2004). Recent trends in the pharmacotherapy of personality disorders. Journal of Personality Disorders 18(1):90-101.

Markovitz PJ and Wagner SC (2003). Efficacy of modafinil augmentation in depression with or without personality disorders. Journal of Clinical Psychopharmacology 23:207-209.

Mattaini M (1999). Clinical Interventions with Families. Washington DC: NASW Press.

Mauri MC, Rudelli R, Somaschini E, Roncoroni L, Papa R, Mantero M, Longhini M, and Penati G (1996). Neurobiological and psychopharmacological basis in the therapy of bulimia and anorexia. Progress in Neuropsychopharmacology and Biological Psychiatry 20(2):207-240.

Miller BC (1995). Characteristics of effective day treatment programming for persons with borderline personaliut disorder. Psychiatric Services 46(6):605-608.

Muller RJ (1992). Is there a neural basis for borderline splitting? Comprehensive Psychiatry 33(2):92-104.

Muller-Oerlinghausen B, Roggenbach J, and Franke L,2004). Serotonergic platelet markers of suicidal behavior—do they really exist? Journal of Affective Disorders 79(1-3):13-24.

New AS, Buchsbaum MS, Hazlett EA, Goodman M, Koenigsberg HW, Lo J, Iskander L, Newmark R, Brand J, O'Flynn K, and Siever LJ.(2004). Fluoxetine increases relative metabolic rate in prefrontal cortex in impulsive aggression. Psychopharmacology (Berl). May 25 E-pub.

Palmer RL, Birchall H, Damani S, Gatward N, McGrain L, and Parker L (2003). A dialectical behavior therapy program for people with an eating disorder and borderline personality disorder–description and outcome. International Journal of Eating Disorders 33(3):281-26.

Palombo J (1983). Borderline conditions: a perspective from self-psychology. Clinical Social Work Journal 11(4):323-338.

Paris J (1997). Childhood trauma as an etiological factor in the personality disorders. Journal of Personality Disorders 11(1):34-49.

Paris J (1993). The treatment of borderline personality disorder in light of the research on its long-term outcomes. Canadian Journal of Psychiatry, 38, Feb., Suppl. 1, 528-534.

Paris J and Zweig-Frank H (1997). Dissociation in patients with borderline personality disorder. American Journal of Psychiatry 154(1):137-138.

Paris J, Zweig-Frank H, Kin NM, Schwartz G, Steiger H, and Nair NP (2004). Neurobiological correlates of diagnosis and underlying traits in patients with borderline personality disorder compared with normal controls. Psychiatry Research 121(3):239-52.

Pennington BF (2002). The Development of Psychopathology: Nature and Nurture. New York: Guilford Press.

Phillips MI and Sierra M (2003). Depersonalization disorder: a functional neuroanatomical perspective. Stress 6(3):157-165.

Pinto OC and Akiskal HS (1998). Lamotrigine as a promising approach to borderline personality: an open case series without concurrent major mood disorder. Journal of Affective Disorders 51: 333-343.

Revenson TA and Seidman E (2002). Looking backward and moving forward: Reflections on a quarter century of community psychology. In Revenson TA, D'Augelli AR, French SE, Hughes DL and Livert D, Handbook of Community Psychology. pp 3-31. New York: Kluwer Academic/Plenum Puboishers.

Rilling JK, Winslow JT, O'Brien D, Gutman DA, Hoffman JM, and Kilts CD (2001). Neural correlates of maternal separation in rhesus monkeys. Biological Psychiatry 49(2):156-67.

Rinne T, van den Brink W, Wouters L, and van Dyck R.(2002). SSRI treatment of borderline personality disorder: a randomized, placebo-controlled clinical trial for female patients with borderline personality disorder. American Journal of Psychiatry 159(12):2048-54.

Rizvi ST (2002). Lamotrigine and borderline personality disorder. Journal of Child and Adolescent Psychopharmacology 12:356-365.

Robins CJ and Chapman AL (2004). Dialectical Behavior Therapy: Current status, recent developments, and future directions. Journal of Personality Disorders 18(1):73-89.

Robins CJ and Koons CR (2004). Dialectical behavior therapy for severe personality disorders. In JJ Magnavita, Ed. Treating Personality Disorders: Theory, Practice, and Research, pp 117-139. New York: Guilford.

Russ MJ, Campbell SS, Kakuma T, Harrison K, and Zanine E (1999). EEG theta activity and pain insensitivity in self-injurious borderline patients. Psychiatry Research 89(3):201-14.

Sadock BJ and Sadock VA (2003). Synopsis of Psychiatry, 9th ed. Philadelphia: Lippincott, Williams and Wilkins.

Schlosberg A (1976). Some factors and issues concerning community psychiatry. International Journal of Social Psychiatry 22(2):120-9.

Schmahl CG, McGlashan TH, and Bremner JD (2002). Neurobiological correlates of borderline personality disorder. Psychopharmacology Bulletin 36(2):69-87.

Schwartz JM, Stoessel PW, Baxter LR jr, Martin KM, and Phelps ME (1996). Systematic change in cerebral glucose metabolic rate after successful behavior modification treatment of obsessive-compulsive disorder. Archives of General Psychiatry 53(2):109-13..

Silk KR (2000). Borderline personality disorder: overview of biological factors. Psychiatric Clinics of North America 23:61-75.

Singh N (2000). Reappraising assessment. Keynote address, Conference on Family Support and Children's Mental Health, Research and Training Center, Portland State University, Portland , Oregon, April.

Smail D (1999). Power, the environment and community psychology: The influence of power on psychological functioning: Community psychology perspectives. Journal of Community and Applied Social Psychology 9(2):75-78.

Springer T and Silk KR (1996). A review of inpatient group therapy for borderline personality disorder. Harvard Review of Psychiatry 3(5):268-278.

Stalenheim EC, von Knorring L and Oreland L (1997). Platelet monoamine oxidase activity as a biological marker in a Swedish forensic psychiatric population. Psychiatry Research 69(2-3):79-87.

Stanley B, Molcho A, Stanley M, Winchel R, Gameroff MJ, Parsons B, and Mann JJ (2000). Association of aggressive behavior with altered serotonergic function in patients who are not suicidal. American Journal of Psychiatry 157:609-614.

Stein DJ, Hollander E, Cohen L, Frenkel M, Saoud JB, DeCaria C, Aronowitz B, Levin A, Liebowitz MR and Cohen L (1993). Neuropsychiatric impairment in impulsive personality disorders. Psychiatry Research 48(3):257-66.

Steinberg BJ, Trestman R, Mitropoulou V, Serby M. Silverman J, Coccaro E, Weston S, de Vegvar M, and Siever LJ (1997). Depressive response to physostigmine challenge in borderline personality disorder patients. Neuropsychopharmacology 17(4):264-273.

Stone MH (1994). Characterologic subtypes of the borderline personality disorder. With a note on prognostic factors. Psychiatric Clinics of North America 17(4):773-84.

Stone MH (1987). Psychotherapy of borderline patients in light of long-term follow-up. Bulletin of the Menninger Clinic 51(3):231-47.

Strohman RC (2003). Genetic determinism as a failing paradigm in biology and medicine: implications for health and wellness. Journal of Social Work Education 39(2):169-191.

Terkelsen K (1983). Schizophrenia and the family: II. Adverse effects of family therapy. Family Process 22:191-200.

Torgersen S, Kringlen E, and Cramer V (2001). The prevalence of personality disorders in a community sample. Archives of General Psychiatry 58(6):590-96.

US Department of Health and Human Services (1999). Mental Health: A Report of the Surgeon General. Dept of Public Health, Department of Health and Human Services.

Verkes RJ, Fekkes D, Zwinderman AH, Hengeveld MW, Van der Mast RC, Tuyl JP, Kerkhof AJ, Van Kempen GM (1997). Platelet serotonin and [3H]paroxetine binding correlate with recurrence of suicidal behavior Psychopharmacology (Berl) 132(1):89-94.

Verkes RJ, Pijl H, Meinders AE, and Van Kempen GM (1996). Borderline personality, impulsiveness, and platelet monoamine measures in bulimia nervosa and recurrent suicidal behavior. Biological Psychiatry

40(3):173-80.

Waldinger, R (1987). Intensive psychodynamic therapy with borderline clients: an overview. American Journal of Psychiatry 144(3):267-274.

Wilson BDM, Hayes E, Greene GJ, Kelly JG, and Iscoe (2003). Community psychology. In DK Freedheim, ed. Handbook of Psychology: History of Psychology, vol. 1 pp 431-449. New York: John Wiley and Sons.

Woods WJ (1975). Community psychiatry. Australian Nurses Journal 4(10):28-31.

Zanarini MC (2004). Update on pharmacotherapy of borderline personality disorder. Current Psychiatry Reports 6(1):66-70.

Zanarini MC and Frankenburg FR (2003). Omega-3 fatty acid treatment of women with borderline personality disorder: A double-blind placebo-controlled pilot study. American Journal of Psychiatry 160:167-169.

Zanarini MC, Ruser T, Frankenburg FR, and Hennen J (2000). The dissociative experiences of borderline patients. Comprehensive Psychiatry 41(3):223-3

Figure 14

Loss of brain volume associated with schizophrenia is clearly shown by magnetic resonance imaging (MRI) scans comparing the size of ventricles (butterfly- shaped, fluid-filled spaces in the midbrain) of identical twins, one of whom has schizophrenia (rights). The ventricles of the twin with schizophrenia are larger. This suggests structural brain changes associated with the illness.

Source: Daniel Weinberger, NIMH Clinical Brain Disorders Branch, National Institute of Mental Health

Combined 3-D PET and MRI scans show human brain activity (red structure at 4:00, just above ear) triggered by clozapine. This is a different pattern than is seen with traditional neuroleptics like haloperidol.

Source: John Hsiao, NIMH Experimental Therapeutics Branch National Institute of Mental Health

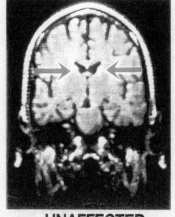
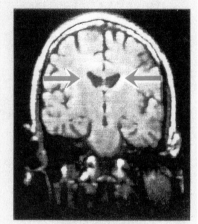

SCHIZOPHRENIA IN MONOZYGOTIC TWINS

Pair no. 2: 44 year old males

UNAFFECTED   AFFECTED

figure 15

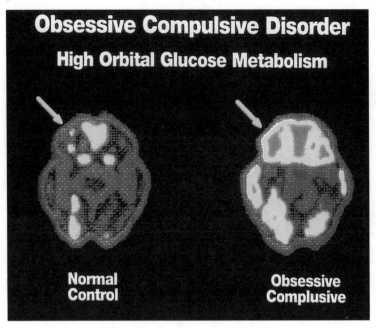

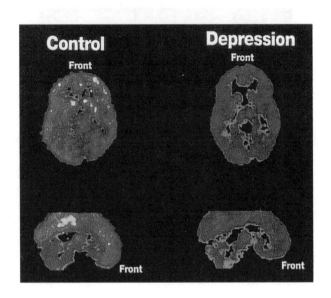

Figure 15

PET (positron emission tomography) scans show increased brain metabolism (brighter colors), particularly in the frontal cortex, in the brain of a person with obsessive-compulsive disorder (OCD) juxtaposed with a brain of a person without OCD. This suggests altered brain function in people with OCD.

Source: Lewis Baxter, University of Alabama

PET (positron emission tomography) scans of the brain of a person without depression (L) and the brain of a person with depression (R) reveal reduced brain activity (darker colors) during depression, especially in the frontal cortex.

Source: Mark George, NIMH Biological Psychiatry Branch
National Institute of Mental Health

This material was made possible by the campaign to End Discrimination Founding Sponsors: Abbott Laboratories, Bristol-Myers Squibb, Eli Lilly, Janssen Pharmaceuticals, Pfizer, Sandoz Pharmaceuticals, SmithKline Beecham, Wyeth-Ayerst Laboratories. In accordance with NAMI policy, acceptance of funds does not imply endorsement of any business practice or product. For more information, contact the National Alliance for the Mentally Ill, http://www.nami.org/

PART V
SUBSTANCE ABUSE
AND ADDICTION

28
How Do We Become Addicted? From Risk Factors to a Changed Brain*

Thou hast the keys of Paradise, oh just, subtle, and mighty opium.

Thomas De Quincey,
Confessions of an English Opium Eater, 1821

What can science teach us about addictive drugs and addictive behavior? That requires a thorough analysis, drug by drug, of how each one acts and what harm each one does to users and to society.

Avram Goldstein,
Addiction from Biology to Drug Policy, 2001, p 13

What is addiction? It's been defined in several ways. We saw in Part IV that definitions of mental illness are strongly influenced by culture; so too are definitions of addiction. Although the World Health Organization (1992) and the American Psychiatric Association (2000) prefer the term dependence, the word addiction is still widely used today. With respect to substances of abuse, addiction is typically defined as compulsive, persistent seeking and using a substance despite major adverse consequences. Addiction to nonconsumables (such as gambling or sexual partners) is similarly understood as compulsive and

---

*I'm indebted to Dr Steven E Hyman, provost, Harvard University, director of the National Institute of Mental Health 1996-2001, and formerly director of Harvard's Interdisciplinary Biobehavioral Program, for permission to include material from a memorable lecture to clinicians in 1994. In an era of amazing advances in the science of addiction, the principles he enunciated a decade ago remain valid today.

persistent seeking and behavior in the face of serious harm to self or others. That is, addiction connotes not only a subjective state of urgent wanting but also behavior, for the purpose of experiencing either intense pleasure or relief from an aversive state. Substances of abuse and addictive activities act as positive reinforcers (giving a "rush" or euphoria) or as negative reinforcers (offering escape from withdrawal symptoms or painful emotional states).

What images does the word addict bring to mind? Perhaps a street person, unwashed, usually in the inner city, living outside the law, and dark-skinned? The facts belie this stereotype. The vast majority of addicted Americans are Caucasian of all socioeconomic levels who are "heavy users" of caffeine, nicotine, alcohol, or sugar—all legal and all, arguably, addictive. Obesity/overweight, now epidemic in the United States, is often a result of one kind of addiction—to the pleasures of taste.

Another way of thinking about addiction is a *vulnerable individual's response* to taking addictive substances with adequate *dose* (enough of it); *frequency* (often enough); and *chronicity* (over a long enough period of time) to enter an addicted state (Hyman, 1995).

Vulnerability is a very complicated concept. The same individual can have different levels of vulnerability in different environments or life situations at different times in his or her life. For example, you may not be a vulnerable person most of the time, but when you lose your significant other or get fired from your job, you may become vulnerable to substance abuse or addiction.

How many of us can claim honestly that we're free of addiction? For example, I know I'm addicted to at least two substances—sugar (I can't conceive of a day without several sweets) and caffeine (which has sustained me through the writing of this book). How am I different from poor homeless people addicted to crack or heroin? Answer: I've had a luckier roll of the dice in life, and my addictions are socially acceptable in the dominant culture. My addictions, however, are similar to those of people addicted to crack or heroin with respect to the neurobiological process by which I became addicted, the eager seeking and compulsive use of the substances, and strong cravings to the point of obsession when the substance is withdrawn, not to mention headache, fatigue, and a depressed mood. Are there serious negative consequences? Though not visible to others, fairly often my body communicates unpleasant sensations related to overconsumption, and there are some ominous warnings. These substances contribute to chronic gastritis and inflammation that are known to predispose to stomach cancer.

Concepts related to addiction are tolerance, dependence, and sensitization. *Tolerance* refers to the loss of effect of a drug after repeated administration over a long time on a frequent schedule, so that more and more is needed to produce the same high.

Dependence is defined as a state in which stopping a drug suddenly ("cold turkey") causes withdrawal sickness which is dramatically relieved by another dose of the same drug (Goldstein, 2001, pp 88-90). Where does the term "cold turkey" come from? The gooseflesh symptom of withdrawal from opiates gave rise to that expression!

Sensitization is the opposite of tolerance. The drug's effect is enhanced rather than diminished with repeated use, which sometimes happens with a few particular drugs when taken rapidly and repeatedly such as cocaine (Goldstein, 2001). Sensitization refers to persistent hypersensitivity to a drug's effect in a person with a history of exposure to that drug (Cami and Farré, 2003).

From risk factors for addiction to a changed brain. In Part I, we defined risk and protective factors. Risk factors are biological or nonbiological variables in individuals and in environments, interacting through time, to cause or exacerbate problems of health, mental health, and social conditions. Protective factors are the opposite of risk factors, acting to diminish problems in these areas. With respect to addiction, the more risk factors and the fewer protective factors we have, the more likely it is that we'll get hooked. These risk factors fall into the categories of individual vulnerability, environmental factors, and drug effects, which differ according to the drug (see Table 6).

Table 6
Risk Factors for Addiction

Individual vulnerability (some aspects can change over time)
- genetic
- psychiatric condition
- chronic pain
- feeling stressed
- user goals (such as experimentation, escape).

Environmental factors
- drug availability—if you can't get it, you won't become addicted no matter how vulnerable you are.
- peer group pressure to use.
- lack of behavioral alternatives to drug use (no opportunities for fun or satisfaction).
- settings in which drugs are used such as religious ceremonies, family holidays.
- presence of conditioned cues (such as being at a place where you used to use frequently, or running into drug-using friends).

Drug effects (drugs differ)
- drug's addictiveness (some are highly addictive, such as cocaine; others are not very addictive, such as LSD).
- drug purity.
- route of administration (for example, you'll get addicted to cocaine faster if you freebase than if you snort.)
- dose.
- frequency of use.
- chronicity of use.

Chronic use of drugs causes long-lived molecular changes in the signaling properties of neurons. Depending on the drug and the circuits involved, these adaptations have different effects on behavior and different time-courses of initiation and decay.

With chronic drug use, three types of changes may take place in the brain centers that control *somatic functions* (body functions), *rewards* and *pleasures*, and *emotional memories*.

We can see physical effects of drugs that affect somatic functions when the drug is withdrawn. We can't see the physical effects of drugs on reward and pleasure pathways in the brain or on emotional memories, but these effects are physical too. They are just as real, and just as physical, as the somatic effects of withdrawal from alcohol (e.g. tremor, hypertension, grand mal seizures, tachycardia, irritability, delusions, hallucinations), caffeine (e.g. headache, fatigue), or opiates (e.g. severe muscle cramps, bone ache, diarrhea, tearing, hypothermia or hyperthermia, insomnia, restlessness, nausea, gooseflesh).

Only a few drugs involve somatic dependence, but almost all drugs of abuse are believed to induce the other two kinds of changes in brain structures and functions. With respect to brain reward and pleasure pathways, changes both in microanatomic structures and chemical processes involve motivation and volition. Motivational aspects of withdrawal are:

- dysphoria—feeling sad, blue, down in the dumps.
- anhedonia—can't experience pleasure. Things you used to enjoy aren't fun anymore.
- cravings—have to have it, fast.

The person suffering from these feelings experiences a change in behavioral priorities. Now, getting the drug of abuse often becomes the most important goal in life.

Changes in *emotional memories* are a hallmark of addiction. During a lifetime, many memories are eventually lost through decay of memory traces in the brain. However, memories of powerful experiences remain. Cues evoke these memories of either intensely pleasurable experiences, leading to cravings in addiction, or intensely painful experiences, leading to traumatic flooding, as in posttraumatic stress disorder (PTSD). These memories are referred to as "privileged" memories because they take precedence both in affecting the individual's emotional state and in motivating behavior.

These latter two types of long-term changes—changes in structures and functions affecting motivation and volition, and changes in emotional memories—are actual physical effects in the brain that you cannot see. We used to distinguish between "physical" and "psychological" addiction by the presence or absence of somatic withdrawal, such as tremors, nausea, or muscle cramps. Now we know that *all drugs of abuse produce actual physical changes in the brain*. Even though many of the physical effects of drugs on the brain are not directly observable, they are just as real as tremors and muscle cramps.

How long do these brain changes last? Somatic withdrawal may last days, weeks, sometimes even longer. Motivational aspects of withdrawal may last from several weeks to months, even years. Emotional memories may last a lifetime—we may never shake them off. That is, once addicted, you may never "withdraw" from the memories of intense pleasure associated with drug use. That's why AA members with years of sobriety call themselves "recovering," not "recovered," alcoholics.

Each drug has its own special neurotransmitter. Drugs of abuse work in the brain through different neurotransmitter systems. Some drugs affect several transmitters through a chain of reactions. Here are some of the transmitters that are active with different drugs:

DRUG	NEUROTRANSMITTER
Opiates/Heroin	Endorphins/enkephalins
Cocaine/Amphetamine	Dopamine
Nicotine	Acetylcholine
Alcohol	GABA, opioids, and others
Marijuana (a cannabinoid)	THC receptor ligand anandamide
Hallucinogens	Serotonin
Caffeine	Adenosine

But.... most roads lead to the dopamine high-way!

260

Hyman (1995) described the above process a decade ago, and research today supports this view (Hyman and Malenka, 2001). The complex neurobiological processes identified over the past decade that underlie these effects are reported in numerous sources (see, for example, Yang, Zheng, Wang et al, 2004; Thompson, Swant, Gosnell et al, 2004; Bolanos and Nestler, 2004; Wang, Gao, Zhang et al, 2003; NIDA, 2004). Wang and colleagues refer to the process described by Hyman as "abnormal engagement of long-term associative memory."

Current theories of addiction rely heavily on neurobiological evidence showing connections between addiction-related behaviors and neural structures and functions. These connections have been identified by imaging, biochemical analyses, genetic studies, and laboratory experiments.

It is widely agreed that all drugs of abuse act on dopamine systems either directly or indirectly. Determining the role of dopamine has been the predominant focus of addiction research during the past 20 years (Kalivas, 2004). Drug seeking and drug self-administering in humans and animals can be triggered either by direct exposure to drugs of abuse or by stressful events, both of which increase strength of excitatory synapses on mesolimbic dopamine neurons (Saal, Dong, Bonci et al, 2003). Moreover, it appears that other forms of addiction, such as compulsive gambling, also create feelings of pleasure, excitement, or satisfaction through dopamine pathways (Cami and Farré, 2003; Goudriaan, Oosterlaan, de Beurs et al, 2004; Ibanez, Blanco, de Castro et al, 2003).

The concept of *reward* is central to most views of addiction. Several brain circuits, structures, and neurotransmitters are involved in the reward process. Dopamine has had front-runner status in this regard for several decades, but other neurotransmitter systems (those that process serotonin, norepinephrine, opioids, GABA, and glutamate) also have important roles in the regulation of reward (see Chapters 10-12 for information about the major neurotransmitters). Among neuroscientists, it has been widely accepted for some time that dopamine mediates the rewarding (reinforcing) properties of natural stimuli such as food and sex as well as drugs of abuse.

However, it appears that the relationship between dopamine and food or sex is more complex than was previously thought. Recent studies indicate that pleasure from food or sex ("hedonic response" in scientific terms) may continue in laboratory animals even when dopamine functions are suppressed. For some addictions, dopamine may promote effort to get and consume (food) or to perform an act (having sex) without mediating the reward process itself (another transmitter may be responsible for reward). Neuroscientist passions run high on both sides of the Great Dopamine Debate: Is dopamine the happy chemical? To find the answer, read the following authors, or if that's too much trouble, you could take my own opinion on faith. I vote for "often, yes, always, no!" (Cannon and Bseikri, 2004; Salamone, Correa, Mingote et al, 2003; Berridge and Robinson, 2003; Wise, 2004; Hajnal, Smith, and Norgren, 2004; Giuliano and Allard, 2001; Paredes and Ågmo, 2004).

Cannon and Bseikri (2004), despite their own research showing that dopamine is not *necessary* for food pleasure in laboratory animals, conclude that there's reason to think it's *important*. In a recent review article on drug addiction, Cami and Farré state unequivocally, "Both natural rewards (food, drink, and sex) and addictive drugs stimulate the release of dopamine" (2003, p 980). Studies have found the transmitter to be involved in drugs of abuse, gambling (Goudriaan et al, 2004), food (Wise, 2004; Hajnal et al, 2004), sex (Giuliano and Allard, 2001), pair-bond formation (Young, Lim, Gingrich et al, 2001), listening to music (Sutoo and Akiyama, 2004), seeing attractive faces (Kampe, Frith, Dolan et al, 2001), video games (Koepp, Gunn, Lawrence et al, 1998), positive social interactions (Vandenschuren, Niesink, Van et al, 1997; Hansen, Bergvall and Nyiredi, 1993), and best of all, humor (Mobbs, Greicius, Abdel-Azim et al, 2003).

Yes, humor is fun and feels good—is it surprising that Mobbs and colleagues, using fMRI (functional magnetic resonance imaging), found that humor activates the dopamine-processing pleasure pathway in the brain? In answer to the question "What is the role of dopamine in reward?" Cannon and Bseikri (2004), citing Mobbs et al, quip that scientific progress would be greatly expedited if we were all simply funnier. I agree! Next assignment: Tell your boss that you were late to work because of a herniated hippocampus.

Other views of addiction expand existing theories of reward by separating its psychological components into *learning* through the experience of using, *liking* that experience (pleasure that diminishes over time), and *wanting* to repeat it (Berridge and Robinson, 2003). These researchers believe that although liking decreases over time, wanting increases, continuing to be a powerful motivator even when using is no longer enjoyable because the brain has become *hypersensitized* to effects of the drug in the context of impaired cognitions that lead to compulsive seeking and using.

These cognitive effects have been connected to events in the frontal cortex. Goldstein and Volkow (2002) have expanded the focus on limbic subcortical structures to include structures in the frontal cortex using findings from neuroimaging studies. They found that the orbitofrontal cortex and the anterior cingulate, regions neuroanatomically connected with limbic structures, are the frontal cortical areas most frequently implicated in drug addiction. These structures are activated in addicted subjects during intoxication, craving, and bingeing; deactivated during withdrawal; and also involved in higher-order cognitive and motivational functions. That is, addiction connotes cognitive and emotional processes, regulated by the frontal cortex, which result in overvaluing drug reinforcers, decreasing sensitivity to alternative reinforcers (that is, perceiving them as less desirable), and deficient inhibitory control for drug responses (Volkow, Fowler, and Wang, 2002). These changes in addiction, called *salience attribution* and *impaired response inhibition,* expand traditional concepts of drug dependence that emphasize limbic system responses to pleasure and reward.

A related view cites evidence that compulsive drug use and its persistence arise from pathological usurpation of molecular mechanisms involved in memory (Hyman and Malenka, 2001), a process we spelled out earlier. This process characterizes the takeover of the addicted person's life so familiar to users, their families, and those of us who try to help them grapple with addiction.

Another theory proposes that specific brain reward and stress circuits become dysregulated during the development of alcohol dependence (Koob, 2003). Parts of the amygdala and the nucleus accumbens (a grouping called the "extended amygdala") mediate multiple neurotransmitter systems that process GABA, opioid peptides, glutamate, serotonin, and dopamine. Withdrawal from drugs of abuse is associated with subjective negative affect, accompanied by action of stress hormones (see Chapter 19). This "toxic" effect of chronic drug use creates an ongoing state of vulnerability for relapse.

There is increasing evidence that the excitatory transmitter glutamate also plays a central role in processes underlying the development and maintenance of addiction (Kalivas, 2004; Tzschentke and Schmidt, 2003). Glutamate dispatched from the prefrontal cortex to the nucleus accumbens appears to promote reinstatement of drug-seeking behavior. That is, glutamate may be a major contributor to relapse. In particular, context-specific aspects of behavior (control over behavior by conditioned stimuli) seem to depend heavily on transmission of glutamate.

The influence of genetics has emerged in conjunction with the publication of the Human Genome study in 2003. Roles of genes in addiction are elusive because substance abuse is a product of polygenetic action and is strongly influenced by environments through epigenetic systems in the brain (see Chapter 7). Data have been available for many years showing much higher risks for alcohol addiction in biological children of substance abusing parents adopted at birth into non-substance abusing homes, than for children without a family history of addiction (Schuckit, 1985). Recently, specific genes and variants of genes (alleles) have been identified that either protect individuals from addiction by, for example, causing aversive effects from ingesting certain substances, or put individuals at risk for various types of substance abuse (for a review of this literature, see Cami and Farré, 2003; Comings and Blum, 2000).

As Cami and Farré (2003) have pointed out, these various theories overlap in some ways and are not mutually exclusive. None alone can explain all facets of addiction. In Parts I-IV we've considered the mind-boggling intricacy of the brain, the most complex entity in the known universe. Substance abuse and addiction happen through interaction between that entity and the world around it, so it's not surprising that these phenomena are extremely complex as well. Little by little, we're piecing together various aspects that give us guidance to work with people.

Popular psychology has expanded the usage of the word "addiction" to include "workaholics" (who work too long and too hard), women who "love too much" (have a pattern of attachments to hurtful partners), and "codependents" (significant others of addicted persons who supposedly have a need to have an addicted partner) (Whitfield, 1991). These groups are thought to have addictive characteristics because they have seemingly overpowering cravings that they compulsively and repetitively seek to satisfy, despite very adverse consequences.

If the underlying neurobiology of attachment to compulsive work or hurtful partners is found to resemble the process that has been identified for addictive drugs and gambling, that interpretation might be justified. However, to the best of my knowledge no such evidence has been uncovered. There are possible alternative explanations for these behavior patterns, for example, anxiety about not making enough money to meet expenses (leading to compulsive working), or an attachment to an addicted partner because of that person's other qualities *despite* the addiction, not *because of* it. In this case, the therapeutic issue is not to determine the partner's "codependency"—connoting that the partner of the addicted person has a sickness equivalent to that of the addicted person—but rather to weigh costs and benefits of the relationship as it is, and to consider strategies for bringing about change.

A body of published critiques of the codependency concept has noted its pejorative connotations, characterizing interpersonal behaviors as addictions or diseases, pathologizing women, promulgating a value-laden Anglo cultural narrative, and requiring partners of addicted persons to assume responsibility for their partner's addiction (Montgomery, 2001; Anderson, 1994; Collins, 1993; Inclan and Hernandez, 1992). Although recent validation studies of instruments measuring problems and issues of family members have shown reliability and validity with respect to characteristics on the inventories (Dear, 2004), the designation "codependent" remains offensive. Studies that simply gather data about the kinds of challenges experienced by families with a member experiencing addiction, in the context of a stress and coping conceptual framework, can supply information that may help practitioners support families (Hurcom, Copelle, and Orford, 2000). Unlike the codependency frame, a stress and coping model normalizes individuals and their behaviors.

The issue of *enabling* (significant others doing things that inadvertently support rather than discourage the addicted person's habit) is separate from the concept of codependency. Enabling does not assume pathology in significant others and is an important target for educational and therapeutic efforts (Rotunda, West, and O'Farrell, 2004).

Brain structures and systems involved in substance abuse and addiction. Brain structures referred to here are described in Chapters 10 and 11. Two systems, the *mesolimbic dopamine pathway*, also called the pleasure pathway, and the *mesocortical dopamine circuit*, originate in neurons in the *ventral tegmental area* (VTA), where neurons manufacture dopamine (see figures 16 and 17 pp 268-269). All drugs of abuse act on these systems at different levels. The two systems act in parallel and also interact with each other to mediate the addiction process.

The mesolimbic dopamine pathway is probably the most important circuit involved in reward. In this system, a bundle of nerve fibers (axons) project from the VTA to the *nucleus accumbens* (NAc) in the limbic system bordering on the basal ganglia. Messages are relayed through the *amygdala* as part of this system. Another limbic structure involved in addiction is the *hippocampus*, which is involved in short-term memory formation. Opioid pathways also participate in reward.

The mesolimbic dopamine pathway has a role in creating privileged memories of highly rewarding novel stimuli. These memories cause addicts to have cravings and risk relapse even after years of abstinence. The limbic system plays a key role in determining what is salient enough to be remembered.

The mesocortical dopamine system projects from the VTA to the prefrontal cortex, the anterior cingulate (mediates response inhibition and initiation), and the orbitofrontal cortex (mediates ability to evaluate future consequences and balance immediate rewards against long-term negative consequences). These structures are involved in the conscious experience of drug-taking, cravings, and compulsions (Cavedini, Riboldi, Keller et al, 2002).

Basal ganglia structures, the *caudate nucleus* and the *putamen* (together called the *striatum*) are also involved. The *hypothalamus,* located near the limbic region and the basal ganglia, is activated when stress is contributing to addictive responses. The hypothalamus controls many hormones, including those that help the individual cope with stress, such as cortisol (see Chapter 19).

The *prefrontal cortex*, critical for higher cognitive functions such as executive planning, working memory, hypothesis generation, response inhibition, action initiation, and problem solving, is also importantly involved in substance abuse and addiction. Some cognitive dysfunctions seen fairly often in heavy users such as poor appraisal of likely consequences of behavior, overvaluation of drug effects, difficulty making decisions, and poor response inhibition, suggest interactions between mesolimbic structures and prefrontal and other cortical structures that mediate cognitive functions.

Figure 17 (p 269) shows some of the routes for neural mes-

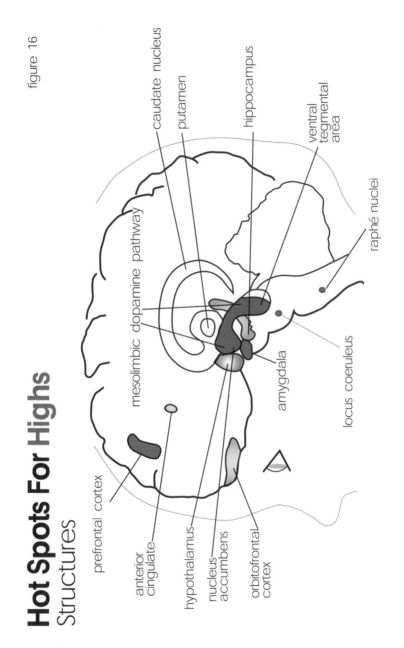

figure 16

Hot Spots For Highs
Structures

caudate nucleus

putamen

hippocampus

ventral tegmental area

raphé nuclei

mesolimbic dopamine pathway

amygdala

locus coeruleus

prefrontal cortex

anterior cingulate

hypothalamus

nucleus accumbens

orbitofrontal cortex

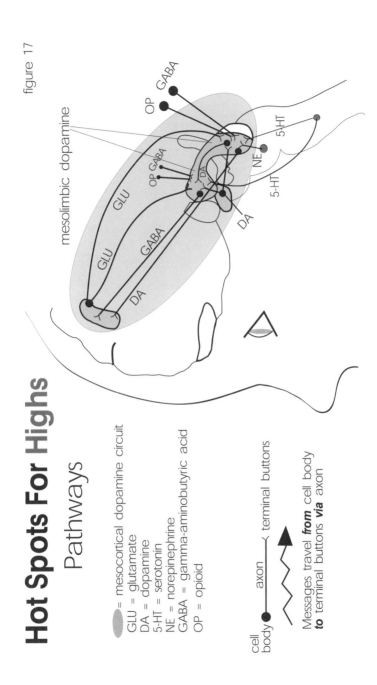

figure 17

Hot Spots For Highs

Pathways

= mesocortical dopamine circuit

GLU = glutamate
DA = dopamine
5-HT = serotonin
NE = norepinephrine
GABA = gamma-aminobutyric acid
OP = opioid

cell body — axon — terminal buttons

Messages travel *from* cell body *to* terminal buttons *via* axon

mesolimbic dopamine

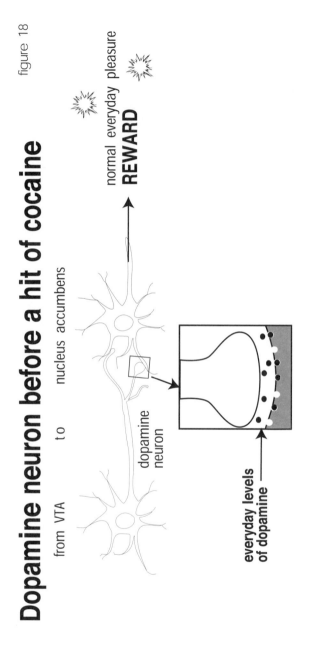

figure 18

Dopamine neuron before a hit of cocaine

from VTA   to   nucleus accumbens

dopamine
neuron

everyday levels
of dopamine

normal everyday pleasure

REWARD

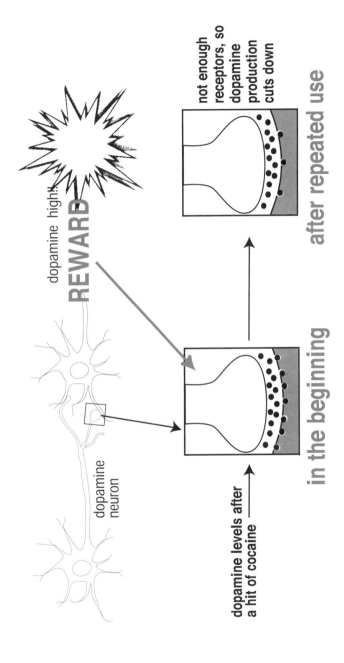

figure 19

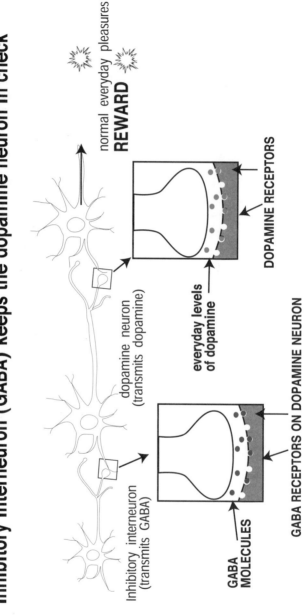

figure 20

Inhibitory interneuron (GABA) keeps the dopamine neuron in check

normal everyday pleasures
REWARD

DOPAMINE RECEPTORS

everyday levels
of dopamine

dopamine neuron
(transmits dopamine)

GABA RECEPTORS ON DOPAMINE NEURON

Inhibitory interneuron
(transmits GABA)

GABA
MOLECULES

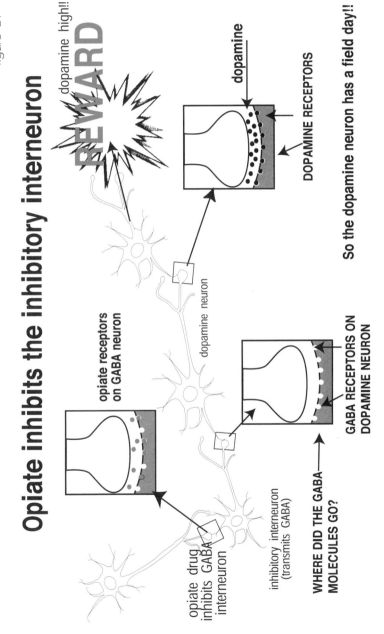

figure 21

Opiate inhibits the inhibitory interneuron

dopamine high!!

REWARD

dopamine

DOPAMINE RECEPTORS

So the dopamine neuron has a field day!!

opiate receptors
on GABA neuron

dopamine neuron

GABA RECEPTORS ON
DOPAMINE NEURON

opiate drug
inhibits GABA
interneuron

inhibitory interneuron
(transmits GABA)

WHERE DID THE GABA
MOLECULES GO?

sages related to reward and cognition..

Let's try to follow these arrows to help us visualize these routes. Dopamine (DA) neurons originate in the VTA and project their axons to the nucleus accumbens, the amygdala, and the prefrontal cortex. Neurons processing glutamate (GLU), the major excitatory transmitter in the brain, send messages from the prefrontal cortex to the VTA and the nucleus accumbens. Neurons processing GABA, the major inhibitory transmitter in the brain, send messages from the nucleus accumbens to the prefrontal cortex and also act on the VTA and the nucleus accumbens. Opioid-processing neurons (OP on the diagram) modulate GABA's inhibitory influence on the release of dopamine in the VTA and also affect the release of norepinephrine (NE) from the locus coeruleus. Serotonin neurons (5-HT) originate in the raphé nuclei and project to the VTA, the nucleus accumbens, and the striatum, where they modulate release of dopamine. These processes are described by Cami and Farré (2003).

These networks are very complicated, and it's not necessary to "master" this information in order to use it to help your clients. You do need a basic understanding of the ways some rewarding and cognitive events take place in relation to specific classes of drugs such as cocaine, opiates, and nicotine, and the ways using drugs or engaging in addictive behaviors leads us to become addicted. We'll explain that later in this chapter.

The prominent role of neurobiology in serious medical, psychological, and social problems related to addiction has led to efforts to identify and use pharmacological agents (i.e. more drugs) to treat drug addiction. Later in this chapter we'll consider the pros and cons of this direction in addiction treatment. First, we'll briefly review the kinds of medications in use.

There are at least three types of drugs for treating addictions: agonists, antagonists, and aversive agents. *Agonists* bind with receptor molecules in a similar fashion to the drug of abuse, prevent withdrawal symptoms, but do not give the high of the drug of abuse. For example, methadone binds to opiate receptors in place of heroin, prevents heroin withdrawal, and takes away the craving for heroin. Nevertheless, the street marketability of some treatment drugs such as methadone suggests that they do give some kind of a high. Another problem with agonists is that they are also likely to be addictive, so the user substitutes one addiction for another. The advantages are that the person may be able to function (e.g., hold a job) better than when using the original drug, and that the prevention of withdrawal symptoms and cravings diminishes the need for criminal behavior to get money for the drug of abuse.

Antagonists bind with receptors in a different way from the drug of abuse (competitive binding at the receptor site), so taking the drug of abuse gives no high because the antagonist drug now occupies the receptors. The craving, however, is not satisfied. For example, naltrexone binds to opiate receptors, blocks the effects of heroin, but does not remove the craving for heroin and other opiates. Since the antagonist therapeutic drug now blocks the effect of the original drug, the user may just substitute a different drug of abuse for the original drug. Researchers are developing and testing new drugs that combine agonist and antagonist actions in the hope of remedying the limitations of each type of medication.

Aversive agents, such as disulfiram (Antabuse) for alcohol addiction, deter drug use by making you very sick if you use the drug while taking the medication.

29
Some Street Drugs--How They Can Get You Hooked

We'll look briefly at the actions of some classes of drugs. Let's start with *stimulants.*

Cocaine, amphetamine, and other stimulants produce euphoria and increase arousal, alertness, concentration, and motor activity. They also increase blood pressure and pulse rate and stimulate the release of stress hormones such as cortisol. With prolonged use, they can cause aggressive behavior, irritability, stereotyped behavior, and paranoia. In fact, people brought to emergency rooms in a state of acute psychosis are sometimes misdiagnosed as having schizophrenia when they are actually experiencing the effects of stimulant overdose.

Stimulants not only affect dopamine neurons directly but also act indirectly by altering the excitability of dopamine neurons through pathways that transmit norepinephrine and glutamate (Paladini, Mitchell, Williams et al, 2004).

Cocaine. Cocaine blocks the reuptake into the presynaptic neuron of three kinds of transmitters, dopamine, norepinephrine, and serotonin, by jumping on board the shuttles that carry transmitters back up into the sending neuron. The word shuttle is a metaphor for molecules, called *transporters*, that are embedded in the presynaptic membrane. Transporters pump neurotransmitters back into presynaptic neurons. Cocaine's main effect works through dopamine and dopamine transporters.

Cocaine grabs the slots on the shuttle away from the dopamine neurotransmitters that are floating around in the synapse waiting to be reuptaken—so there's a lot more dopamine left in the synapse than before the person had a hit of cocaine. That is, cocaine makes a direct hit on the dopamine pathways.

Look at figure 18 (p 270). The bundles of dopamine neurons that form the mesolimbic dopamine pathway extend from the VTA in the midbrain to the nucleus accumbens in the limbic system. The diagram shows the ordinary everyday dopamine neuron. When you indulge in your usual pleasures, like eating a bagel with cream cheese or listening to music, you get a little shot of dopamine. If you're lucky, you get a lot of little shots during a typical day. The bagel and the music give you those little shots.

Figure 19 (p 271) shows what happens when you take a hit of cocaine. The cocaine blocks the normal reuptake of some of the dopamine molecules into the presynaptic neuron. This means that a lot of dopamine is now available to combine with the postsynaptic receptor molecules that are located on the surface of the receiving neuron. Now your synapses are flooded with your own dopamine.  The dopamine is what gives you the rush. *You get high on your own dopamine!*

What happens when continuing floods of dopamine triggered by cocaine sledgehammer the dopamine system? The brain tries to adapt to this sledgehammering. After you've been using cocaine at a sufficient dose, often enough, over a long enough period of time, DNA (genetic material) is transcribed (read out to make a copy) as RNA, which is then translated into protein (see Chapter 7). This protein is a precursor of the neurotransmitter dynorphin, which tells neurons to cut down on dopamine production. The brain's adaptation to too much dopamine is to reduce the number of dopamine receptors. So the brain adapts to the presence of cocaine, and when you take the cocaine away, you're in an adapted state.

Now you're in trouble. You don't have enough dopamine receptors, so you have symptoms of *dysphoria* (depression) and *anhedonia* (inability to experience pleasure). All this leads to *cravings.* You're much worse off than you were before you took the drug because you don't feel just a sort of normal blah, now you feel severe psychic pain. This psychic pain is neurobiological in origin. It comes from molecular changes in the brain brought about by genetic activity in response to drug use.

New research is giving evidence of cocaine-induced dysregulation of glutamate systems. Glutamate mechanisms are now recognized as underlying several clinical aspects of cocaine dependence, including euphoria, withdrawal, craving, and pleasure dysfunction (Dackis and O'Brien, 2003). Researchers think that even denial, traditionally assumed to be purely psychological, appears to result in part from glutamate dysfunction in cortical regions. Hm·mm....could that mean that when I believe (erroneously) that I have plenty of money in my checking account, it's just my glutamate gone awry? Medications have not as yet been effective for treating cocaine addiction, but at least two novel treatments are being tried (Dackis and O'Brien, 2003). A promising drug for treating cocaine-addicted individuals by enhancing glutamate function is modafinil, which we encountered earlier as a treatment for clinically depressed people with borderline personality disorder who had sleeping spells (narcolepsy).

The most radical idea is to treat cocaine addiction with viruses. The organisms in question display cocaine-binding proteins on their surfaces that can sequester cocaine in the brain, thereby attenuating the effects of cocaine on the user (Carrera, Kaufmann, Mee et al, 2004). An attractive feature of this approach is that the viruses have no pharmacodynamic action of their own and therefore have little potential for side effects. (But would you want to make room for a few more creatures to live in your body?)

Amphetamine makes a direct hit on dopamine systems as well, but it entails a more complicated process. It causes vesicles storing dopamine to release dopamine faster into the synapse from the presynaptic neuron. Like cocaine, it also inhibits reuptake of dopamine, norepinephrine, and serotonin by membrane transporters. The net result, again, is a lot more dopamine in the synapse available to combine with postsynaptic receptors. Once again, the rush comes from our own dopamine. Different mechanism, same effect.

Like cocaine, amphetamine has recently been found to increase dopamine release in the nucleus accumbens through a mechanism regulated by glutamate transmission from the prefrontal cortex (Ventura, Alcaro, Mandolesi et al, 2004).

In the case of "ecstasy" (MDMA), a so-called "designer derivative " of amphetamine, the serotonin transporter is the main target. Ecstasy gives euphoria and increased empathy (called an "entactogenic" effect). Ecstasy and other designer derivatives of amphetamine can have toxic effects on dopamine and serotonin neurons. Acute intoxication with psychostimulants can cause cerebral hemorrhage, hyperthermia, heat stroke, panic, psychosis, or "serotonin syndrome" (altered mental state, neuromuscular abnormalities).

Damage from methamphetamine use has now been found to be even worse than previously thought (Thompson, Hayashi, Simon et al, 2004). In a study comparing 22 long-term users with 21 controls matched for age, the users had lost an average of 11% of tissue in the limbic system and were depressed, anxious, and unable to concentrate. The hippocampus, the brain's center for making new memories, had lost 8%, a deficit comparable to brain deficits in early Alzheimer's.

Opiates. Opiate drugs such as heroin and morphine give euphoria, sedation, and calmness. However, with repeated use, tolerance and strong physical dependence occur. Overdose can cause fatal respiratory depression.

Opiates' effects are produced through a chain of actions using the transmitter GABA (gamma-aminobutyric acid). Remember that neurons come in different sizes and shapes, and that each neuron can connect with many other neurons. Short neurons make connections with longer neurons. These interconnecting short neurons are called interneurons. There are lots of interneurons in the brain that transmit an inhibitory neurotransmitter called GABA (gamma-aminobutyric acid). These inhibitory interneurons inhibit the action of receiving neurons. These are the neurons that just say no!

Look at figure 20 (p 272). Here you see a dopamine neuron in its everyday state, giving little ordinary pleasures. The short GABA interneuron is keeping the dopamine neuron in check by sending it some inhibitory messages. The GABA interneurons send inhibitory messages to receiving neurons and balance out the messages these receiving neurons are getting at the same time from excitatory neurotransmitters. When the right amount of GABA is coming through, you get modest amounts of dopamine—just the right amount of dopamine to enjoy your bagel with cream cheese or your music.

But what happens when the GABA neurons stop squirting GABA onto the dopamine neurons? Not enough inhibitory messages? Right! The balance between excitatory and inhibitory messages is disrupted. The dopamine neuron gets overexcited and fires faster and faster.

That's what happens when opiate drugs get into the system. Now look at Figure 21 (p 273). Opiates (like heroin and morphine) also have inhibitory effects. They inhibit the inhibitory GABA neurons (that's a double negative). Opiates say no to GABA neurons, so the GABA neurons *stop saying no* to the dopamine neurons. Did you notice in the diagram that the neurotransmitter molecules being sent from the GABA interneuron to the dopamine neuron are gone? That's because the opiate drug has inhibited the GABA neuron. Result: you get a flood of dopamine.

The dopamine neurons have a field day!

Alcohol (ethanol). Alcohol is often referred to as ethanol in the research literature to identify it as the kind of alcohol you drink to have fun. We'll use the terms interchangeably here. Almost immediately after a couple of drinks (especially if you drink on an empty stomach), alcohol gives a buzz or even a little euphoria as the drug stimulates release of dopamine from mesolimbic dopamine neurons. Then you start to feel relaxed as well. It acts on receptors for GABA, an inhibitory transmitter, and potentiates (strengthens) GABA's action of making you relaxed, mellowed out, maybe even a little silly.

But are you one of those people who reacts quite differently to alcohol? Do you get angry and aggressive? For many years, "mad drunks" were assumed to be releasing pent-up anger that they'd been suppressing because of the disinhibiting effects of alcohol. Now, however, we know that disinhibition—

which allows us to express underlying feelings we were afraid to express when sober—is not the only explanation for why good-hearted Glenn has rages when he's drunk. Chemicals in the drink can actually create a rage in the brain that was not present before using the drug, because of the interaction between that person's limbic chemistry and the drug's properties. It's possible that your friend Glenn, who's always so nice except when he's drunk, isn't covering anything up when he's sober—he's really nice and he isn't the least bit angry. When he's drunk, though, he becomes nasty. He isn't just letting his "true colors" show—the chemistry in his brain has actually been altered by the drug.

Your other friend Beverly, however, also gets nasty when intoxicated—she's the wife of a misogynist and gets punished for being too assertive. In her case, that anger really *was* already present and really *did* come out because of the disinhibiting properties of the three martinis she had at your party (but who's counting?).

Glenn and Beverly illustrate the great diversity of the brain's responses to environmental inputs. Gerald Edelman, Nobel laureate and neuroscientist, assures us that no two people, and no two experiences shared by those people, are ever exactly the same, even for identical twins and even from the day of birth (well, earlier, actually—situations in utero aren't identical either). This is so because epigenesis continuously creates differences in gene expression (Edelman, 2004) (see Chapter 22).

Alcohol consumption changes concentrations of several neurotransmitters (dopamine, GABA, endogenous opioid peptides, serotonin, glutamate, and noradrenaline). These changes are associated with activation of reward centers in the brain, disinhibition, and "mellowing out." Dopamine and norepinephrine mechanisms, together with endogenous opioid peptides, are thought to play important roles in the reinforcing properties of alcohol through activation of positive reinforcement pathways.

Alcohol potentiates GABA (causing relaxation) and stimulates release of dopamine from mesolimbic neurons (causing euphoria). It also reduces the firing rate of pars reticulata (PR) neurons, which are believed to have an inhibitory effect on dopamine neurons, so ethanol disinhibits the dopamine neurons through an indirect process. Low doses of ethanol produce significant enhancement of the release of dopamine in limbic structures in mice.

Currently, three drugs are approved for treatment of alcohol addiction: disulfiram, acamprosate, and naltrexone. Another drug, ondansetron, a serotonin-3 receptor antagonist, is being tested and has been found in early trials to be effective in early-onset alcoholism but not late-onset (Kranzler, Pierucci-Lagha Feinn et al, 2003). People with early-onset alcohol dependence have a greater familial loading for alcoholism, more severe progression of the disorder, a greater severity of additional psychiatric disorders, and a poorer response to treatment than those with late-onset alcoholism.

There is still no evidence from randomized controlled clinical trials that disulfiram improves abstinence rates over the long term (Mann, 2004), although individuals have given testimonials that it has helped them. Also, aversive therapy with disulfiram entails risks for some users and requires careful monitoring.

Acamprosate and naltrexone have improved outcomes for many alcohol-dependent clients. In a meta-analysis of all studies through 2001, Kranzler and Van Kirk (2001) found that both naltrexone and acamprosate are efficacious in reducing alcohol consumption in persons addicted to alcohol.

Acamprosate was found extremely effective in 13 of 16 European trials (Soyka and Chick, 2003) but not effective in a single controlled six-week trial in the United States (Brasser, McCaul, and Houtsmuller, 2004). The action of acamprosate is unclear, but for many alcohol-dependent people, it has worked to reduce relapse.

Naltrexone is an opioid antagonist that blocks mu-opioid receptors. This action may produce its anti-relapse effects either by blocking opioids' reinforcing effects when alcohol is consumed or by diminishing conditioned expectation of these effects (Littleton and Zieglgansberger, 2003). It is still not clear whether naltrexone's major effect is to reduce cravings or to prevent the pleasurable effects of alcohol.

Opioid peptide antagonists, drugs that block the effects of opioid peptide neurotransmitters, decrease self-administration of ethanol by laboratory animals. Early clinical trials have shown that these drugs can decrease relapse rates in detoxified outpatient alcoholics (Fuller and Gordis, 2001). Some recovering alcoholics report that opioid antagonists such as naltrexone take away their cravings for alcohol, even though these same opioid blockers don't take away cravings for opiate drugs. In fact, in the early 1900s, morphine, an opiate drug, was used to treat alcoholism!

However, recent research has shown mixed results. Naltrexone was no more effective than placebo in a large multisite study of veterans, almost all male, who had been severely addicted for decades (Krystal, Cramer, Krol et al, 2001). In their analysis of the work of Krystal and colleagues in light of the overall history of naltrexone trials, Fuller and Gordis (2001) agree with the caveats presented by the authors that the lack of effectiveness for the population of alcohol-addicted older male veterans may differ from results for younger persons with shorter histories of use, involving more females, and in social environments more conducive to hope for the future. The psychosocial treatment in the veterans' study consisted only of outpatient counseling 1 hour per week for the first 16 weeks, then less often. Better results have been obtained when naltrexone is combined with much more intensive interpersonal treatments. Thus, despite recent negative results, tests of naltrexone as a treatment for alcohol addiction continue with a range of populations.

Nicotine. The World Health Organization estimates that 1 billion smokers worldwide smoke 6 trillion cigarettes per year, causing the death of more than 3 million persons each year (Sadock and Sadock, 2003). Nicotine binds to receptors for the neurotransmitter acetylcholine, causing an increase in number and sensitivity of acetylcholine receptors. This process probably accounts for withdrawal symptoms when the drug is withheld, because there are now too many unfilled receptors. Nicotine produces reinforcing effects by acting on the nicotinic subtype of acetylcholine receptors located on dopamine neurons in the VTA, stimulating the release of dopamine to the nucleus accumbens and the cerebral cortex. It may also sensitize nicotinic acetylcholine receptors located on glutamate terminals. It enhances glutamate's excitation of dopamine neurons and decreases GABA's inhibition of dopamine neurons, leading possibly to long-lasting heightened sensitivity of dopamine neurons (Pidoplichko, Noguchi, Areola et al, 2004).

Caffeine. Caffeine is an antagonist to a transmitter that blocks dopamine D1 receptors, called adenosine. It acts as another of those double negatives. Caffeine blocks the dopamine blocker, thereby enhancing dopamine levels. Caffeine induces dopamine-dependent behavioral arousal and enhances motor activity in rodents. It also activates norepinephrine-producing neurons. Prolonged use of caffeine can lead to physical dependence as evidenced by withdrawal symptoms during abstinence. Some tolerance occurs, but in most cases the effects of caffeine continue over time (Nehlig, 2000). Using autoradiography, a procedure that locates radioactive substances in slices of tissue, Nehlig and Boyet (2000) found that caffeine activated multiple systems: caudate nucleus, raphé nuclei, locus coeruleus, striatum, thalamus, ventral tegmental area, and amygdala.

Caffeine was found to have no effect on mood in rested subjects, nor did it improve mood that was depressed due to lack of sleep (James and Gregg, 2004). (Does your personal experience belie this finding? Mine does!)

Brain imaging was used to measure blood volume in the frontal cortex during mental work, both with and without caffeine (Higashi, Sone, Ogawa et al, 2004). The volume of blood decreased during rest periods as a result of caffeine-induced constriction of the cerebral blood vessels but increased during mental work. However, volume increased equally with and without caffeine. Performance on complex mental tasks, however, clearly improved (I knew there must be a good reason not to give up my addiction). This improvement was not reflected in increased blood volume, however. The results indicate that caffeine activates neurons in the prefrontal cortex.

In an experiment with eight habitual coffee drinkers, neuroimaging demonstrated that placebo conditions can induce release of dopamine in the thalamus and the putamen at similar levels to those found with oral caffeine (Kaasinen, Aalto, Nagren et al, 2004). This finding is consistent with other studies showing that placebo effects involve dopamine as well as endogenous opioids (Sher, 2003). (If I believed I was getting caffeine from a look-alike taste-alike stand-in for coffee, could I kick the habit and get the same results? Not likely—placebo effects are often powerful initially, but wear off after a while. Otherwise, maybe we'd have put the pharmaceutical industry out of business by now?).

Marijuana (a cannabinoid). Marijuana's effects include general euphoria, mild release from inhibition, and sometimes sensory perception distortions, stimulated appetite, drowsiness, relief from anxiety, and even altered states of consciousness. Toxic effects include carcinogenesis and cognitive impairment, especially when used together with alcohol (Jacques, Zombek, Guillain et al, 2004). Contrary to previous belief, cannabinoids are now known to trigger reinstatement of drug-seeking behavior in animals previously weaned from intravenous self-administration of drugs (Gardner, 2002). The preponderance of neurobiological studies indicates that cognitive impairments in users are probably reversible, but consequences of use in pregnancy have been serious: long-lasting cognitive alterations in children whose mothers used marijuana during pregnancy (Mereu, Ferraro, Cagiano et al, 2003).

The psychoactive component of marijuana, delta 9-THC (delta 9-tetrahydrocannabinol), appears to stimulate dopamine neurons in the VTA by depressing GABA inhibitory inputs to these neurons (Szabo, Siemes, and Wallmichrath, 2002). Another desired effect may be the extinction of aversive memories, a function mediated by a part of the amygdala known to perform this function. A study of mice treated with the chemical subsequently showed memory impairment as well as elevated levels of natural cannabinoids in that region of the amygdala (Marsicano, Wotjak, Azad et al, 2002).

Cannabinoid agonists' value for multiple medical treatments is best-known with respect to nausea in cancer chemotherapy. Its use is also suggested to stimulate appetite, reduce inflammatory pain, and reduce spasticity and tremor in diseases characterized by motor impairment (multiple sclerosis, spinal cord injury, Huntington's and Parkinson's diseases), as well as for glaucoma, bronchial asthma, and vasodilation that accompanies advanced cirrhosis (Pertwee and Ross, 2002). Combining low doses of cannabinoids and opioids to treat acute and chronic pain offers promise, especially for pain that may be resistant to opioids alone (Cichewicz, 2004). While some of these applications are in the process of testing and most fall in the category of "provisional," they are promising and merit further trials.

Hallucinogens (psychedelics) powerfully alter perception, mood, and cognition. They are not known to produce dependence or addiction. Hallucinogens stimulate the serotonin receptor 5-HT(2A), known to play an essential role in cognitive processing and working memory (Nichols, 2004). Stimulation of these receptors in turn leads to higher levels of glutamate in the cortical regions. These drugs engage the thalamocortical circuits that are central to altered states of consciousness (see Chapter 22), and imaging studies have shown that hallucinogens increase prefrontal cortical metabolism.

The preceding review of several drugs raises a question of burning concern to persons with addictions, their loved ones, and those of us who work to try to help them: *Why do persons with addictions self-destruct?* Steven Hyman expressed the answer powerfully: it's because addiction usurps brain mechanisms that are prime movers in motivated behavior (Hyman, 1995). Addiction commandeers the parts of the brain that control motivation and set behavioral priorities. The addict is only aware of dysphoria and craving, and believes that everything would be OK with just another hit or another drink. "If only I could get another hit off my crack pipe, I'd be able to talk to you about treatment." The "strange" behavior of an addicted person—who despite the disasters that have resulted from drug use, still sees the drug as the most important thing in life—seems not so strange when we realize that the person's brain is really changed.

30
How Do Psychosocial Interventions Work in the Context of a Changed Brain?

Interventions for substance abuse include support groups such as AA, NA, and smoke-enders; cognitive-behavioral strategies; residential communities; other psychosocial interventions; and medication.

Like a person who has had a stroke, an addicted person has lost certain nerve cells but can recover some functions because other parts of the brain are sometimes able to take over what the damaged part of the brain used to do. Psychological, social, and contextual treatments support this process. They act as prostheses for the person's broken brain. By asking people to take responsibility for themselves, you're asking for other thinking parts of the brain and other emotional parts of the brain to do things the damaged part of the brain used to do.

It doesn't make the changes go away. Problems are still there, especially long-term emotional memories that predispose people to relapse. But psychosocial treatments enlist other parts of the brain to act as bulwarks against these parts of the brain that want nothing more than to get another hit. Similar principles work in trauma as well. You can never get rid of the traumatic emotional memories, but you can get other parts of the brain to manage. You learn ways of suppressing and managing these memories, focusing on other things, and utilizing strategies to circumvent the force of those memories. But the memories are always there.

But treat drug abuse with drugs? That sounds like craziness!
Maybe so. But often it helps, in conjunction with psychosocial interventions. In each individual case, we need to ask (1) what are the risks of our proposed interventions? (2) what are the likely benefits of our proposed interventions? and (3) what are the possible risks of *not doing* the proposed interventions? That third question (which is often sadly neglected) is really on target with respect to treating drug abuse with drugs. If we had a drug that could take away cravings and help the person stay clean or sober, and it was a cliffhanger between relapse and sobriety, wouldn't there be a risk attached to not giving that medication?

Successful treatment requires that the addicted person eventually take personal responsibility for himself or herself, overcome denial, commit to sobriety, and take active steps to achieve it. Once a commitment is made, we can then provide ways of propping up the compromised brain. The same principles apply in most serious chronic diseases. The individual is asked to comply with treatment and to avoid behaviors that put him or her at high risk of relapse or of worsening the condition. For example, a person is at high risk for a heart attack because he or she is obese, has a family history of heart disease, smokes, and is a couch potato. Only he or she can actually change that situation by following the doctor's orders to lose weight, change eating habits, stop smoking, and exercise. The person with the disease must take responsibility for compliance. Just because you were unlucky enough to have been more at risk than some other people doesn't change the fact that only you can behave in ways that will perpetuate or discourage the condition.

A major goal of research on addiction is to identify the molecular basis for long-lasting changes in behavior induced by drugs of abuse. In a fascinating study of the patterns of actions of different drugs in mouse brains, all drugs of abuse strongly activated a particular enzyme, kinase, in the nucleus accumbens, lateral bed nucleus of the stria terminalis, central amygdala, and deep layers of the prefrontal cortex, working through a dopamine D1 receptor mechanism. Although some non-addictive drugs activated kinase mildly in a few of these areas, they never induced a strong pattern of activation. Antidepressants and caffeine activated kinase in the hippocampus and cerebral cortex (Valjent, Pages, Herve et al, 2004).

Does this study suggest that a biochemical intervention might prevent or reverse addiction or dependence for *all drugs of abuse*, with just one or two magic bullets? No one knows just yet, but this result, first published in April 2004, will undoubtedly become a focus of ongoing study, and no doubt, controversy as well.

References Part V

American Psychiatric Association (2000). Diagnostic and Statistical Manual of Mental Disorders, 4th ed, text revision. Washington DC: American Psychiatric Association.

Anderson SC (1994). A critical analysis of the concept of codependency. Social Work 39(6):677-85.

Aragona BJ, Liu Y, Curtis JT, Stephan FK, and Wang Z (2003). A critical role for nucleus accumbens dopamine in partner-preference formation in male prairie voles. Journal of Neuroscience 23(8):3483-90.

Berridge KC and Robinson TE (2003). Parsing reward. Trends in neuroscience 26(9):507-13.

Blood AJ and Zatorre RJ (2001). Intensely pleasurable responses to music correlate with activity in brain regions implicated in reward and emotion. Proceedings of the National Academy of Sciences 98(2):11818-23.

Bolanos CA and Nestler EJ (2004). Neurotrophic mechanisms in drug addiction. Neuromolecular Medicine 5(1):69-83.

Brasser SM, McCaul ME, and Houtsmuller EJ (2004). Alcohol effects during acamprosate treatment: A dose-response study. Alcohol Clinical and Experimental Research 28(7):1074-1083.

Cami J and Farré M (2003). Mechanisms of disease: Drug addiction. New England Journal of Medicine 349(10):975-986.

Cannon CM and Bseikri MR (2004). Is dopamine required for natural reward? Physiology and Behavior 81(5):741-48.

Carlson NR (2001). Physiology of Behavior, 7th ed. Boston: Allyn and Bacon.

Carrera MR, Kaufmann GF, Mee JM, Meijler MM, Koob GF, and Janda KD (2004). From the cover: Treating cocaine addiction with viruses. Proceedings of the National Academy of Sciences USA 101(28):10416-21.

Cavedini P, Riboldi G, Keller R, D'Annucci A, and Bellodi L (2002). Frontal lobe dysfunction in pathological gambling patients. Biological Psychiatry 51(4):334-41.

Cichewicz DL (2004). Synergistic interactions between cannabinoid and opioid analgesics. Life Sciences 74(11):1317-24.

Collins BG (1993). Reconstructing codependency using self-in-relation the ory: a feminist perspective. Social Work 38(4):470-76.

Comings DE and Blum K (2000). Reward deficiency syndrome: genetic aspects of behavioral disorders. Progress in Brain Research 126:325-41.

Dackis C and O'Brien C. Glutamatergic agents for cocaine dependence. Annals of the New York Academy of Sciences 1003:328-45.

Dear GE (2004). Test-retest reliability of the Holyoake Codependency Index with Australian students. Psychological Reports 94(2):482-4.

De Quincey T (1971). Confessions of an English Opium Eater. New York: Viking Penguin.1971. First published in 1821. Quoted in Goldstein, 2001 p 159.

Edelman GM (2004). Wider than the Sky. New Haven: Yale University Press.

Fuller RK and Gordis E (2001). Naltrexone treatment for alcohol depend ence. New England Journal of Medicine 345:1770-71.

Gardner EL (2002). Addictive potential of cannabinoids: the underlying neu robiology. Chemistry and Physics of Lipids 121(1-2):267-90.

Giuliano F and Allard J (2001). Dopamine and male sexual function. European Urology 40(6):601-608.

Goldstein A (2001). Addiction from Biology to Policy, 2nd ed. New York: Oxford University Press.

Goldstein RZ and Volkow ND (2002). Drug addiction and its underlying neurobiological basis: neuroimaging evidence for the involvement of the frontal cortex. American Journal of Psychiatry 159:1642-52.

Goudriaan AE, Oosterlaan J, de Beurs E, and Van den Brink W (2004). Pathological gambling: a comprehensive review of biobehavioral find ings. Neuroscience Biobehavioral Review 28(2):123-41.

Hajnal A, Smith GP, and Norgren R (2004). Oral sucrose stimulation increases accumbens dopamine in the rat. American Journal of Physiology-Regulatory Integrative and Comparative Physiology 286(1):R13; 286(1):R31-7.

Hansen S, Bergvall AH, and Nyiredi S (1993). Interaction with pups enhances dopamine release in the ventral striatum of maternal rats: a microdialysis study. Pharmacology and Biochemistry of Behavior 45(3):673-6.

Higashi T, Sone Y, Ogawa K, Kitamura YT, Saiki K, Sagawa S, Yanagida T, and Seiyama A (2004). Changes in regional cerebral blood volume in frontal cortex during mental work with and without caffeine intake: functional monitoring using near-infrared spectroscopy. Journal of Biomedical Opt. 9(4):788-93.

Hoffman, PL and Tabakoff, B (1996). Alcohol dependence: a commentary on mechanisms. Alcohol and Alcoholism 31(4):333-340.

Hurcom C, Copello A, and Orford J (2000). The family and alcohol: effects of excessive drinking and conceptualizations of spouses over recent decades. Substance Use and Misuse 35(4):473-502.

Hyman SE (1995). What is addiction? Harvard Medical Alumni Review, Winter.

Hyman SE and Malenka RC (2001). Addiction and the brain: The neurobiology of compulsion and its persistence. Nature Reviews Neuroscience 2(10):695-703.

Ibanez A, Blanco C, de Castro IP, Fernandez-Piqueras J, Saiz-Ruiz J. (2003). Genetics of pathological gambling. Journal of Gambling Studies 19(1):11-22.

Inclan J and Hernandez M (1992). Cross-cultural perspectives and codependence: The case of poor Hispanics. American Journal of Orthopsychiatry 62(2):245-55.

Iversen L (2001). Drugs: A Very Short Introduction. New York: Oxford University Press.

Jacques JP, Zombek S, Guillain Ch, and Duez P (2004). Cannabis: Experts agree more than they admit. Revue Medicale de Bruxelles 25(2):87-92.

James JE and Gregg ME (2004). Effects of dietary caffeine on mood when rested and sleep restricted. Human Psychopharmacology 19(5):333-41.

Kaasinen V, Aalto S, Nagren K, and Rinne JO (2004). Expectation of caffeine induces dopaminergic responses in humans. European Journal of Neuroscience 19(8):2352-6.

Kalivas PW (2004). Glutamate systems in cocaine addiction. Current Opinions in Pharmacology 4(1):23-9.

Kampe KK, Frith CD, Dolan RJ, and Frith U (2001). Reward value of attractiveness and gaze. Nature 413(6856):589.

Koepp MJ, Gunn RN, Lawrence AD, Cunningham VJ, Dagher A, Jones T, Brooks DJ, Bench CJ, and Grasby PM. (1998). Evidence for striatal dopamine release during a video game. Nature 393(6682):266-268.

Koob GF (2003). Neuroadaptive mechanisms of addiction: Studies on the extended amygdala. European Neuropsychopharmacology 13(6):4422-52.

Kranzler HR, Pierucci-Lagha A, Feinn R, and Hernandez-Avila C (2003). Effects of ondansetron in early- versus late-onset alcoholics: A prospective, open-label study. Alcohol Clinical and Experimental Research 27(7):1150-5.

Kranzler HR and Van Kirk J (2001). Efficacy of naltrexone and acamprosate for alcoholism treatment: a meta-analysis. Alcohol Clinical and Experimental Research 25(9):1335-41.

Krystal JH, Cramer JA, Krol WF, Kirk GF and Rosenheck RA (2001). Naltrexone in the treatment of alcohol dependence. New England Journal of Medicine 345:1734-9.

Littleton J and Zieglgansberger W (2003). Pharmacological mechanisms of naltrexone and acamprosate in the prevention of relapse in alcohol dependence. American Journal of Addiction 12 Suppl 1:S3-11.

Lopez HH and Ettenberg E (2002). Sexually conditioned incentives: Attenuation of motivational impact during dopamine receptor antagonism. Pharmacology and Biochemistry of Behavior 72(1):65-72.

Malcolm RJ (2003). GABA systems, benzodiazepenes, and substance dependence. Journal of Clinical Psychiatry 64 Suppl 3:36-40.

Mann K (2004). Pharmacotherapy of alcohol dependence: a review of the clinical data. CNS Drugs 18(8):485-504.

Marsicano G, Wotjak CT, Azad SC, Bisogno T, Rammes G, Cascio MG, Hermann H, Tang J, Hofmann C, Zieglgansberger W, Di Marzo V, and Lutz B (2002). The endogenous cannabinoid system controls extinction of aversive memories. Nature 418(6897):530-4.

Mereu G, Fa M, Ferraro L, Cagiano R, Antonelli T, Tattoli M, Ghiglieri V, Tanganelli S, Gessa GL and Cuomo V (2003). Prenatal exposure to a cannabinoid agonist produces memory deficits linked to dysfunction in hippocampal long-term potentiation and glutamate release. Proceedings of the National Academy of Sciences USA 100(8):4915-20.

Mobbs D, Greicius MD, Abdel-Azim E, Menon V, and Reiss AL (2003). Humor modulates the mesolimbic reward centers. Neuron 40(5):1041-8.

Montgomery H (2001). Codependency through a feminist lens. Praxis 1:59-65.

Nehlig A (2000). Are we dependent upon coffee and caffeine? A review on human and animal data. Neuroscience and Biobehavioral Review 23(4):563-76.

Nehlig A and Boyet S (2000). Dose-response study of caffeine effects on cerebral functional activity with a specific focus on dependence. Brain Research 858(1):71-7.

Nestler EJ (2000). The molecular basis of long-term plasticity underlying addiction. National Review of Neuroscience 2:119-28, erratum. National Review of Neuroscience 2:215.

Nichols DE (2004). Hallucinogens. Pharmacological Therapy 101(2):131-81.

NIDA (2004). National Institute on Drug Abuse. See website: http://www.nida.nih.gov/.

Paladini CA, Mitchell JM, Williams JT, and Mark GP (2004). Cocaine self-administration selectively decreases noradrenergic regulation of metabotropic glutamate receptor-mediated inhibition in dopamine neurons. Journal of Neuroscience 24(22):5209-15.

Paredes RG and Ågmo A (2004). Has dopamine a physiological role in the control of sexual behavior? A critical review of the evidence. Progress in Neurobiology 73(3):179-225.

Pertwee RG and Ross RA (2002). Cannabinoid receptors and their ligands. Prostaglandins Leukot Essential Fatty Acids 66(2-3):101-21. Review.

Pidoplichko VI, Noguchi J, Areola OO, Liang Y, Peterson J, Zhang T, and Dani JA (2004). Nicotinic cholinergic synaptic mechanisms in the ventral tegmental area contribute to nicotine addiction. Learning and Memory 11(1):60-9.

Purves D, Augustine GJ, Fitzpatric D, Katz LC, LaMantia AS, McNamara JO, and Williams SM (2001). Neuroscience, 2nd ed. Sunderland, MA: Sinauer Associates.

Rotunda RJ, West L, and O'Farrell TJ (2004). Enabling behavior in a clinical sample of alcohol-dependent clients and their partners. Substance Abuse Treatment 26(4):269-76.

Sadock BJ and Sadock VA (2003). Synopsis of Psychiatry, 9th ed. Philadelphia: Lippincott, Williams and Wilkins.

Saal D, Dong Y, Bonci A, and Malenka RC (2003). Drugs of abuse and stress trigger a common synaptic adaptation in dopamine neurons. Neuron 37(4):577-82.

Salamone J, Correa M, Mingote S, and Weber SM (2003). Nucleus accumbens dopamine and the regulation of effort in food-seeking behavior: Implications for studies of natural motivation, psychiatry, and drug abuse. Journal of Pharmacology and Experimental Therapeutics 305(1):1-8.

Schuckit MA (1985). Genetics and the risk for alcoholism. Journal of the American Medical Association. 254(18):2614-7.

Sher L (2003). The placebo effect on mood and behavior: Possible role of opioid and dopamine modulation of the hypothalamic-pituitary-adrenal system. Forsch Komplementarmed Klass Naturheilkd 10(2):61-8.

Shizgal P and Arvanitogiannis A (2003). Neuroscience. Gambling on dopamine. Science 299(5614):1856-8.

Soyka M and Chick J (2003). Use of acamprosate and opioid antagonists in the treatment of alcohol dependence: a European perspective. American Journal of Addiction 12 Suppl 1:S69-80.

Sutoo D and Akiyama K (2004). Music improves dopaminergic neurotransmission: Demonstration based on the effect of music on blood pressure regulation. Brain Research 1016(2):255-62.

Szabo B, Siemes S, and Wallmichrath (2002). Inhibition of GABAergic neurotransmission in the ventral tegmental area by cannabinoids. European Journal of Neuroscience 15(12):2057-61.

Thompson PM, Hayashi KM, Simon SL, Geaga JA, Hong MS, Sui Y, Lee JY, Toga AW, Ling W, London ED (2004). Structural abnormalities in the brains of human subjects who use methamphetamine. Journal of Neuroscience 24(26):6028-36.

Thompson AM, Swant J, Gosnell BA, and Wagner JJ (2004). Modulation of long-term potentiation in the rat hippocampus following cocaine self-administration. Neuroscience 127(1):177-85

Tzschentke TM and Schmidt WJ (2003). Glutamatergic mechanisms in addiction. Molecular Psychiatry 8(4):373-82.

Valjent E, Pages C, Herve D, Girault JA, and Caboche J (2004). Addictive and non-addictive drugs induce distinct and specific patterns of ERK activation in mouse brain. European Journal of Neuroscience 19(7):1826-36.

Vandenschuren LJ, Niesink RJ, Van J and Ree M (1997). The neurobiology of social play behavior in rats. Neuroscience and Biobehavioral Review 21(3):309-326.

Ventura R, Alcaro A, Mandolesi L, and Puglisi-Allegra S (2004). In vivo evidence that genetic background controls impulse-dependent dopamine release induced by amphetamine in the nucleus accumbens. Journal of Neurochemistry 89(2):494-502.

Volkow ND, Fowler JS, and Wang GJ (2002). Role of dopamine in drug reinforcement and addiction in humans: Results from imaging studies. Behavior and Pharmacology 13(5-6):355-66.

Wang HR, Gao XR, Zhang KG, and Han JS (2003). Current status in drug addiction and addiction memory (abstract). Sheng Li Ke Xue Jin Zhan 34(3):202-6 (article in Chinese).

Wise RA. (2004). Dopamine and food reward: back to the elements. American Journal of Physiology-Regulatory Integrative and Comparative Physiology 286(1):R13.

Whitfield, CL (1991). Co-Dependence: Healing the Human Condition. Deerfield Beach, FL: Health Communications Inc.
World Health Organization (1992). The ICD-10 classification of mental and behavioral disorders: clinical descriptions and diagnostic guide lines. Geneva: WHO.

Yang Y, Zheng X, Wang Y, Cao J, Dong Z, Cai J, Sui N and Xu L (2004). Stress enables synaptic depression in CA1 synapses by acute and chronic morphine: possible mechanisms for corticosterone on opiate addiction. Journal of Neuroscience 24(10):2412-20.

Young LJ, Lim MM, Gingrich B, and Insel TR (2001). Cellular mechanisms of social attachment. Hormones and Behavior 40(2):133-8.

Part VI
Multiple Routes to Quality of Life

Part VI
Multiple Routes to Quality of Life
by Jill V Haga

Multiple routes to quality of life include complementary and alternative medicine, referred to as CAM, and other behaviors and practices usually considered beyond the scope of medicine (such as productive work and exercise). We include this topic in the second edition of *Psyche and Synapse* because it seems important for practitioners to begin to consider these strategies for helping clients as a routine part of the assessment process.

With respect to CAMs, some remedies that now are part of mainstream medical practice were used by practitioners decades or even centuries ago without the benefit of modern science. For example, lithium salts from mineral water springs in early Rome and Greece were prescribed for manic episodes by the fifth century BC physician Aurelianus. Lithium has a strong record of effectiveness in treating bipolar disorder but has been available in America only since 1970—a tragedy for individuals with bipolar disorder and their families, who suffered intensely when American mental health professions knew of no effective help. One of Harriette Johnson's earliest clients and his wife, daughter, and sisters were victims of our ignorance about a resource discovered in a natural source 2,500 years ago. Alas, we learned about lithium too late to have perhaps spared this family the excruciating pain they endured for decades.

Interest in approaches to health and well-being outside the purview of conventional American medicine appears to have snowballed in the past decade and a half, as demonstrated in a recent population survey of American adults by the Centers for

299

Disease Control's (CDC) National Center for Health Statistics (Barnes, Powell-Griner, McFann et al, 2004), and by a recent flurry of scholarly publication.

According to the CDC, during the 12 months prior to the survey 62% of respondents used some form of CAM therapy when the definition of CAM included prayer specifically for health reasons. When prayer specifically for health reasons was excluded from the definition, 36% of adults reported having used CAM therapy during the preceding 12 months. The 10 most commonly used CAM therapies were prayer specifically for one's own health (43.0%), prayer by others for one's own health (24.4%), natural products (18.9%), deep breathing exercises (11.6%), participation in prayer group for one's own health (9.6%), meditation (7.6%), chiropractic care (7.5%), yoga (5.1%), massage (5.0%), and diet-based therapies (3.5%) (Barnes et al, 2004).

An analysis of Medicaid programs revealed that several CAM therapies are now eligible for reimbursement, with 72% of states covering chiropractic treatments, 22% biofeedback, 15% acupuncture, and 11% hypnotherapy and naturopathy (Steyer, Freed, and Lantz, 2002).

CAM use varies by sex, race, geographic region, health insurance status, use of cigarettes or alcohol, and record of hospitalization. CAM was most often used to treat back pain or back problems, head or chest colds, neck pain or neck problems, joint pain or stiffness, and anxiety or depression. Among adults age 18 years or over who used CAM, 54.9% said they believed that CAM combined with conventional medical treatments would help. Most adults who have ever used CAM used it within the 12 month period prior to the survey.

With respect to scholarly writings, we found 49 PubMed entries on touch and well-being, 59 entries on the physiology of touch, 64 empirical studies of mindfulness-based stress reduction, 64 entries for pets and well-being, 92 for pets and emotions, 127 for spirituality and well-being, 865 for spirituality and health, 817 for exercise and well-being, 1,738 for herbal medicine, 9,161 for acupuncture, 10,092 for hypnosis, and

29,739 entries for the single keyword mindfulness, refering to approaches to mental tranquility or fulfilment that entail efforts to pay attention to and be aware of one's thoughts, feelings, and bodily sensations.

Although this body of literature is still very small in comparison with that for traditional medicine, psychiatry, and psychology, it represents exponential growth from the volume of such literature 15 years ago. Only a handful of scholarly articles on any of these topics are listed in major databases prior to 1990. Published works in this area have typically relied mostly on anecdotal accounts. However an increasing number of articles on CAM in scholarly journals are based on well controlled research. It's clear that the healing professions have begun to take a broader view of therapeutic strategies.

The value of touch was discussed briefly in the first edition of this book. In Chapter 31, we'll also consider spirituality, work, exercise, pet therapy, and herbal medicine. All approaches involve the three major dimensions of psychological experience (emotion, behavior, and thought), with these approaches engaging the psyche through differing access routes. Some are initiated mostly by doing, others mostly by thinking or reflecting. Access routes that emphasize doing (behavior) include work, exercise, holding and communicating with pets, taking substances (like herbal remedies), and active interventions from healers (such as acupuncture and hypnosis). Access routes that emphasize thinking and reflecting include spirituality and mindfulness.

No matter what access route is taken, quality of life as a whole is affected—the person's emotions, behavior, cognition, sense of well-being, and neurophysiology are engaged. By now readers are well versed in the integrated mind/body framework of this book, so the concept that any route to our psyches creates ripple effects in all areas of psychological function should not come as a surprise.

Many of the studies we've reviewed throughout the book have alluded to the continual interactions between neural systems and every other body system, including the immune system and the endocrinological system. New disciplines are emerging within Western medicine that address the holistic nature of illnesses, such as *psychoneuroimmunology* (referring to interactions between psychological, neural, and immune systems), and *psychoneuroendocrinology* (referring to interactions between psychological, neural, and hormonal systems). These interacting body systems are sensitive to environmental influences both positive and negative (see Chapters 19-21).

The delicate balance among functioning systems known as homeostasis can be disrupted by triggers such as disease, stress, or negative emotions. The body tries to regain balance by activating neural and hormonal systems that fortify immune responses (Rabin, 2002). Stress and negative emotions can activate disease, hasten its progression, or inhibit recovery (Cohen, 1996). The immune response applies to a variety of diseases ranging from the common cold to cancer (Kiecolt-Glaser, Robles, and Glaser, 2002). The immune system is particularly influential in the onset and progression of autoimmune diseases such as diabetes or lupus (Cohen, 1996) and immune disorders such as HIV/AIDS (Antoni, 2003). Limbic structures (including the hippocampus, amygdala, and nucleus accumbens) also contribute to activating immune response (Haas and Schauenstein, 1997).

The knowledge that stress and negative emotions contribute to the onset and progression of physical illness raises important issues for conventional medical practitioners. Situations or events that produce stress or negative emotions must be addressed in addition to any medicinal treatment in order to maximize patient recovery. Complementary alternative medicine (CAM) as well as nonmedical pursuits such as spirituality, productive work and exercise can mediate stress and emotional trauma to improve immune system functioning.

Following her review of some of these strategies in Chapter 31, Jill Haga describes the creative use of a few of these approaches to help her son David, diagnosed with autism. She shares David's story with us in the hope that we can help ourselves, our own families and friends, and, if we're practitioners in helping professions, our clients, by expanding the range of possibilities we consider to enhance the quality of life.

Chapter 31
Multiple Routes
Spirituality, Work, Exercise, Pet Therapy, Touch, and Herbal Medicine

Spirituality. Research on the association of spiritual and religious experience with well-being and mental health strongly supports connections between quality of life and spiritual experience in various populations. Most studies have shown that religious involvement and spirituality are associated with better health outcomes, greater longevity, coping skills, health-related quality of life even during terminal illness, and less anxiety, depression, and suicide (Mueller, Plevak, and Rummans, 2001).

Some of the populations for whom spiritual experience was found beneficial were early stage cancer patients; chronically physically ill, terminally ill, and alcohol-dependent persons; women caregivers; male veterans who had experienced sexual assault, elderly Americans; elderly Chinese women; persons with depression; children and adolescents with psychiatric challenges; psychiatric inpatients; and persons with AIDS (Koenig, George, Titus et al, 2004; Zemore and Kaskatus, 2004; Faver, 2004; Chang, Skinner, Zhou et al, 2003; Benjamins, 2004; Coleman, 2004; Mabe and Josephson, 2004; Levin, 2002; Schwartz, Meisenhelder, Ma et al, 2003; Boey, 2003; Schapman and Inderbitzen, 2002; Cohen, 2002; Baetz, Larson, Marcoux et al, 2002; Spiegel and Fawzy, 2002; and Boudreaux, O'Hea, Chasuk, 2002).

Spiritual involvement helps counteract the deleterious effects of stress and negative emotions on the immune system in a number of ways including social support, a sense of control, and increased well-being . Religious people often feel that they can take control of a disease by enlisting the aid of God, or that the presence of the disease is part of a bigger plan rather than some random event that cannot be predicted or controlled. The perceived support and interpersonal contact with an omniscient and omnipresent God, as well as the act of disclosure (such as confessing sins) all contribute to improved immune function (Brown, 2002; Milstein, 2004). Many religions advocate controlling negative emotions (such as anger) and negative attitudes (such as hate or jealousy), thus avoiding chronic emotional states that contribute to immune dysfunction (Williams, 2002).

Increased social support from other members of the religious community also appears to contribute significantly to well-being associated with spirituality. Individuals with a greater amount of social connections and support live longer and happier lives and show less physical pain or psychological symptoms.

Gay and lesbian Christians (n = 23), including four drag queens, insisted on maintaining their identity as Christians despite a widespread discourse that denotes Christian faith as mutually exclusive with homosexual or transgender identities (Sullivan-Blum, 2004). In order to reconcile that view of Christianity with their gender and sexual identities, they developed innovative approaches that the author sees as liberating for people with diverse sexual orientations because their solutions challenge some of Christianity's gender and sexual dichotomies.

Religious families are more likely to use complementary alternative medicine not only in the form of spirituality and faith healing but also massage therapy, herbal medicines, and dietary interventions (McCurdy, Spangler, Wofford, Chauvenet, and McLean, 2003). The people most likely to use faith-based healing are women, African Americans, evangelical Protestants, the elderly and those who are poorer, sicker, and less educated (Mansfield, Mitchell, and King, 2002). Unlike the social support that comes with spirituality and provides an overall sense of well-being (Boudreaux, O'Hea, and Chasuk, 2002), true belief is a necessary component for faith healing and prayer to be effective (Griffiths, 2002). Additionally, many patients want their doctors to consider their spiritual needs as part of treatment; some even want doctors to pray with them (King and Bushwick, 1994).

Research is limited on the neurobiological underpinnings of spirituality—that is, how spiritual experience effects change on the neurobiological level to enhance or diminish quality of life. Using PET scans, Borg, Andree, Soderstrom et al (2003) found that in normal male subjects spirituality is associated with a several-fold variability in serotonin receptor density (5-HT 1A). Comings, Gonzales, Saucier et al (2000) studied a specific gene (DRD4) in 200 male subjects (81 college students and 119 subjects from an addiction treatment unit) and measured personality characteristics using the Temperament and Character Inventory. Strong associations were shown between the presence of alleles of the gene and self-transcendance and spiritual acceptance. The investigators postulated that this effect was a function of the high concentration of dopamine D4 receptors in cortical areas, especially the frontal cortex.

An unusual example of the close ties of religious experience to underlying physiological events is seen in the results of a recent study of the experience of having auras of ecstasy or pleasure (perceptual and emotional signs that a seizure is about to take place) in persons being evaluated for partial epilepsy. Eight of 11 patients had sensory hallucinations, four had erotic sensations, five described a religious or spiritual experience, and several had symptoms judged to have no counterpart in human experience. Eight of the 11 *wanted* to have seizures, five were believed to have induced their own seizure states voluntarily, and four were noncompliant with treatment (Asheim, Hansen and Brodtkorb, 2003). This example is not meant to suggest that most religious experience is related to brain dysfunction—far from it. The overwhelming majority of Americans with no mental differences consider religion a positive part of their lives.

Although the majority of research shows spirituality, prayer and religion as beneficial to individuals for whom it is an important part of life, there are detrimental aspects of religion. The use of prayer and spirituality varies. If there are strong elements of shame and guilt as part of the religion, then there may be adverse psychoneuroimmunological consequences, notably in a rise in levels of neurochemicals that interfere with immune functioning (Dickerson, Kemeny, Aziz, Kim, and Fahey, 2004). From a macro perspective, past history and present political struggles are replete with examples of the misuse of religion on a large scale to dominate and oppress.

Work, employment and volunteering. Employment status can contribute to stress, act as a protective factor, or combine both effects, depending upon individual needs and the type of employment. Employment entails complex issues for some people, such as those entering retirement and mothers receiving welfare.

Society places a premium on being a productive member of society. Unemployment damages self esteem, and the longer the unemployment, the lower the self esteem (Goldsmith, Veum, and Darity, 1997). Being employed offers increased well-being and social status as well as economic benefits (Boardman, Grove, Perkins, and Shepherd, 2003). In some cases, volunteer work can confer benefits similar to those of regular employment. For the mentally challenged or mentally ill, employment can also provide increased health, family satisfaction, safety, and quality of life (Vandenboom and Lustig, 1997) as well as having an impact on the civil rights of these groups (Boardman, Grove, Perkins, and Shepard, 2003).

Employed individuals with HIV reported a higher quality of life than their unemployed counterparts (Blalock, McDaniel, and Farber, 2002). In another study of individuals with suicidal tendencies, full-time employment was a protective factor against suicide (Kraut and Walld, 2003).

Among African American mothers suffering from parental stress and depression, employed mothers were much less likely to spank their children than unemployed mothers on welfare (Jackson, Brooks-Gunn, and Blake, 1998). Going to work is a form of respite from parental duties, possibly making it easier to handle the stress of parenting. However, a study of the impact of welfare-to-work programs found that welfare mothers tend to be placed in low-wage high-stress jobs that may in fact adversely affect the mother-child relationship (Parcel and Menaghan, 1997).

The impact of employment on elderly individuals is also complex. Among elderly men who had experienced devastating personal losses, employment buffered the impact as indicated by improvements in physical but not mental health, both 1 year and 3 years after bereavement (Fitzpatrick and Bosse, 2000). Retirement is a mixed blessing for elderly individuals. Bridge employment after retiring from one's primary job can help some retirees transition from a life of full time employment to one of leisure (Kim and Feldman, 2000).

Volunteering can function as a substitute for employment, help individuals feel useful, and provide a peer group for companionship and support (Barlow and Hainsworth, 2001). Particularly among the elderly, volunteers report higher levels of well-being (Morrow-Howell, Hinterlong, Rozario, and Tang, 2003) and present higher levels of mental health (Schwartz, Meisenhelder, Ma, and Reed, 2003) than nonvolunteers. Elderly volunteers report less pain, an increased desire to participate in life's activities, and a newfound purpose for living (Barlow and Hainsworth, 2001). Volunteering improves social and psychological functioning by countering negative moods and lowering depression among individuals over age 65 (Musick and Wilson, 2003).

Thus employment and volunteering benefit individuals by providing a purpose, social and/or economic benefits, and an overall sense of well-being. However, the benefits depend on the fit between the individual, the activity, and the person's life situation.

Exercise has long been known to provide health benefits and is now known to convey mental health benefits as well. In a review of research on mental health effects of exercise, Callaghan (2004) noted that exercise reduces anxiety, depression, and negative mood and improves self-esteem and cognitive functioning. Exercise is also associated with improvements in the quality of life of those living with schizophrenia. However, despite its well-documented benefits, exercise is seldom recognized by mainstream mental health services as an effective intervention and is consequently underutilized as a mental health intervention (ibid).

In a meta-analysis of studies of the effect of exercise on depression as measured by the scores on the Beck Depression Inventory (BDI), Lawlor and Hopker (2001) found a large decrease in depression symptoms in response to exercise (effect size = 1.1). However, these authors dismissed the result because many of the studies in the meta-analysis used flawed methodology.

A lively debate ensued as other researchers made statistical corrections for methodologically unacceptable studies and still found a robust effect size (.66) (Callaghan, 2004). Using these corrections, depressed persons who exercised were .66 standard deviations less depressed and scored 7.3 points lower on the BDI that those who did not exercise. Critics of Lawlor and Hopker also proposed that they had underestimated the benefits of exercise by excluding several types of studies from their review, notably large population surveys, that indicate strong antidepressant effects of exercise (Biddle, 2001; Mutrie, 2002).

Neurobiological explanations for antidepressant effects of exercise in animal studies have suggested that exercise stimulates secretion of endogenous opioids (Pert and Bowie, 1979) and serotonin (Morgan, 1985), two groups of neurotransmitters known to enhance mood in humans.

Whatever your position in this debate, exercise is relatively free from adverse side effects (except possibly in people with heart-related conditions) and therefore can be tried with very low risk.

Substantial benefits with respect to anxiety have also been

identified in meta-analyses (Petruzzello, 1990) based on data from 104 published and unpublished studies reported between 1960 and 1989. Effects were estimated for state (current) anxiety, trait (dispositional) anxiety, and psychophysiological correlates of anxiety for both short-term intensive and long-term exercise. Effect sizes varied from small to large depending on which measure of anxiety and which type of exercise was involved, ages of the exercisers, and whether or not they had a psychiatric illness. One finding consistent across type of anxiety, exercise, and population was that exercise of more than 20 minutes duration per session is necessary to reduce anxiety levels.

Proposed neurobiological explanations for effects include reducing muscle tension by raising body temperature, a calming effect from release of acetylcholine, and energizing physiologically by increasing adrenaline levels (Callaghan, 2004). Exercise is also thought to distract people from anxiety-provoking situations (Bakre and Morgan, 1978).

A recent prospective randomly assigned study of 82 participants in a 12-week bicycle program or control condition showed positive effects from exercise on depression, anxiety, self-concept, aerobic fitness, heart rate, and maximum oxygen intake not only at the end of the 12 weeks but also on follow-up a year later, effects not seen in control participants (DiLorenzo, Bargman, Stucky-Ropp et al, 1999).

Cognitive functioning too showed advances in response to exercise in two meta-analyses (Van Sickle, Hersen, Simco et al 1996; Etnier, Salazar Landers et al, 1997). Van Sickle and colleagues examined effects of bicycle riding, walking, callisthenics, flex and stretch exercise, aerobic workouts of varying intensity and duration, and nonaerobic stretch and coordination exercises. Outcome measures included performance on tests of memory, mathematics, IQ, and perceptual organization. Exercise produced moderate improvement in cognitive functioning, with yoga producing the largest effect for most outcome measures. In studies with adequate statistical power to give a valid indicator of effects, 47% of subjects showed benefits of exercise on indicators of cognition.

Etnier and colleagues (1997) meta-analyzed 134 studies among participants of all ages (range 6 to 90 years). Overall, the study showed larger effects from long-term exercise than from intensive short-term exercise, and benefits appeared to be more marked in older adults. However, neither of the two meta-analyses could explain what the fairly impressive effect sizes mean in terms of longevity and general quality of life (Callaghan, 2004).

Individual studies have suggested that the more you exercise, the less likely it is you'll have psychological and physical symptoms of stress (Ensel and Nan, 2004). Cancer survivors (Blanchard, Stein, Baker, Dent, Denniston, Courneya, and Nehl, 2004) and those with cardiopulmonary disease (Rojas, Schlight, and Hautzinger, 2003) reported a higher quality of life when they exercised. Efficacy cognition was also directly related to exercise (Oman and Duncan, 1995). Especially for the elderly, exercise can provide self efficacy and social support (McAuley, Jerome, Marquez, Elavsky and Blissmer, 2003), with improvement in functional performance and psychosocial functioning following moderate exercise (Means, O'Sullivan, and Rodell, 2003).

Pet therapy. The introduction of pet therapy to the healing professions gave official recognition to what many have known for centuries, that animal bonds with humans are often sustaining, nurturing, and energizing. In recent years, studies have confirmed that interacting with animals confers many benefits including companionship, exercise, emotional support, tactile comfort, a link with reality, something to care for, and a non-judgmental friend who loves unconditionally. It is particularly beneficial to individuals who are isolated from other humans either by a stigmatizing disability or socially unacceptable behaviors or lifestyles, but also can enhance quality of life for many others without special needs (Brodie and Biley, 1999).

Isolation can have devastating consequences for physical, emotional, and psychological well-being. Depression decreases in dementia patients in response to contact with pets (McVarish, 1995). Terminally ill cancer patients report less fear and anxiety with pet therapy (Muscher, 1984). For hospice patients, pets provide pleasure, stress reduction, and a sense of purpose all of which contribute to the quality of life. They may also become transitional objects, allowing the patients to distance themselves emotionally from humans as part of the dying process and yet still have valuable personal companionship (Fried, 1996). Patients suffering from posttraumatic stress disorder sometimes pull away from humans but can bond with animals (Altschuler, 1999). Elderly homosexuals who feel ostracized by society experience affection, loyalty, acceptance and less loneliness with animals (Kehoe, 1990). In addition to these emotional benefits, pet therapy lowers heart rates and blood pressure (Kaminski, Pellino, and Wish, 2002).

Pets can help severely emotionally disturbed children develop empathy, give them something to touch and experience with affection, and provide unconditional love, acceptance, and trust (Reimer, 1999). Children with a range of handicapping conditions may develop self-esteem and self-efficacy by caring for and loving a pet as they earn recognition and acceptance from nondisabled people in their lives (Ross, 1983). Individuals with communication deficits such autism or aphasia may be particularly drawn to the nonverbal, reassuring and reinforcing nature of animals (Brodie and Biley, 1999).

The physical acts of owning and caring for pets are also beneficial in other ways. Elderly pet owners get more exercise and have higher overall levels of activity than those without pets (Hooker, Holbrook, Freeman et al, 2002). The process of caring for pets has been instrumental in modifying violent tendencies in some prison inmates as they bond not only with the animals but also with fellow inmates and instructors also involved with the animals (Moneymaker and Strimple, 1991).

Human-pet interactions usually involve touching. It seems likely that the frequent physical contact between humans and their pets is a major contributor to the psychological and physical benefits of pet care, although studies have not been able to separate out the relative contributions of different aspects of animal-human interactions.

Touch therapies and massage. Touch as a route to well-being has a long history in Eastern philosophies. Central to these philosophical perspectives of touch are concepts that person and environment are a single entity, that energy flows between them, and that a delicate balance of energy is necessary for healthy functioning. Diseases and disorders arise when the flow of energy is compromised by pathogens or stress. Massage and touch therapies help bring equilibrium back to the energy flow and thus return the person to good health (Bonadonna and Ramita, 2002).

Western medicine is beginning to implement touch-related treatments for a range of conditions. Massage therapy lessens stress and anxiety and helps lower heart rates and blood pressure (Moyer, Rounds, and Hannum, 2004). Both massage therapy and healing-touch therapy lowered blood pressure, respiratory rate, and heart rate in cancer patients (Post-White, Kinney, Savik, Gau, Wilcox, and Lerner, 2003). In a sample of 36 healthy adults, massage decreased stress and increased positive affect (Diego, Field, Sanders, and Hernandez-Reif, 2004).

In the treatment of chronic pain of the head, neck, shoulders, back, and limbs, both conventional treatment and classical massage therapy showed initial improvements. However, only the massage group showed improvement at the follow-up visit three months later (Walach, Guthlin, and Konig, 2003). Patients with musculoskeletal ailments responded to massage with improvements in stress level, pain, sleeping, and coping skills (Weze and Leathard, 2004). Compared to mental relaxation for musculoskeletal pain, massage therapy alleviated muscle pain and enhanced mental energy, and self reported overall health (Hasson, Arnetz, Jelveus, and Edelstam, 2004).

One study indicated that the process of administering massage helps parents bond with their children (Merino-Navarro, Garcia-Melchor, Palomar-Gallardo et al, 2002). Benefits of massage therapy for infants include earlier discharge from the hospital, higher developmental scores, and higher confidence among parents of their parenting skills (Beachy, 2003). Another study revealed that after only five days of massage therapy, preterm infants spent less time sleeping and demonstrated a 53% greater weight gain than similar infants who had not received massage (Deiter, Field, Hernandez-Reif, Emory et al, 2003). This latter reference reports one of several studies by the Touch Research Institutes on benefits of touch for various populations, including premature infants, cocaine-exposed infants, infants parented by depressed mothers, and full-term infants without medical problems. Childhood conditions in which these improvements took place included sexual and physical abuse, asthma, autism, burns, cancer, developmental delays, dermatitis, diabetes, eating disorders, juvenile rheumatoid arthritis, posttraumatic stress disorder, and psychiatric problems. When the massage therapy was given by parents and grandparents, there were added benefits of cost-effectiveness and enhancement of the caregivers' well-being (see references at http://www.miami.edu/touch-research/index.html).

The positive effects of touch demonstrated in these studies suggest that a planned use of touching as a therapeutic strategy may benefit infants, children, and adults experiencing a range of difficult life circumstances. These findings raise an interesting question: is touch alone, even in the absence of increased emotional responsiveness by caretakers, an effective psychotherapeutic technique? If so, it could be that even caretakers whose psychological states impair their abilities to respond proactively, or caretakers beleaguered with job demands (such as being responsible for a large number of elderly persons in a nursing home), can be taught to enhance the well-being of those they care for with frequent and regular touch.

The area of touch warrants much more study as a therapeutic technique. As yet, there are no outcome studies of long-term use of touch therapies with humans. However, given what we have learned about long-term as well as short-term effects of stress on animals, it seems likely that ongoing use of techniques that improve physiological functions related to stress may protect against some of the harmful effects of long-term stress.

Herbal and complementary medicine. As in traditional medicine, studies of effectiveness of various treatments are problematic when they are published only in journals advocating the method being studied and authored by practitioners of that method or approach. We caution readers to consider the sources of the results we report, to judge for themselves the reliability of the sources.

For instance, naturopathic ear drops were found superior to antibiotic treatments for ear infections and had no side effects (Sarrell, Cohen, and Kahan, 2003). Naturopathic medicine has been reported as helpful in treating problems such as asthma, menopause, and otitis media (middle ear infections). Menopausal women reported that naturopathic medicine was equal in effectiveness to conventional medicine in treating anxiety, hot flashes, and vaginal dryness but was 7 times more effective in treating insomnia and low energy (Cramer, Jones, Keenan, and Thompson, 2003). Asthma patients reported increased well-being , freshness, and comfortable breathing following treatments with naturopathy and yoga (Sathyaprabha, Murthy, and Murthy, 2001).

A study of cancer patients showed that people often seek alternative treatments out of fear or uncertainty rather than a true belief in efficacy (van der Zouwe, van Dam, Aaronson, and Hanewald, 1994). However, trust, belief, and expectation are extremely important in positive responses to treatments (Esch, Guarna, Bianchi, Zhu, and Stefano, 2004), a fact consistent with documented large effects of placebo in many illnesses. Complementary alternative medicines work via the frontal, prefrontal, and limbic regions of the brain particularly in the left anterior region involving reward/motivation circuitry. As we have discussed, these regions are believed to be sensitive to positive affect and emotion related memory processing (Esch et al, 2004).

Some herbs have direct effects. In a review article on echinacea, the consensus among studies was that echinacea is effective in reducing the duration and severity of certain symptoms, but that this effect is noted only with certain preparations of the plant. Echinacea and its active components affect specific aspects of the immune system. It has been found effective for reducing the severity and duration of colds, upper respiratory infections, and bronchitis (Percival, 2000).

Diet therapy has been effective for treating characteristics of autism in children. Knivsberg, Reichelt, and Nodland (2001) found that by eliminating foods containing gluten (wheat, oats, rye, and barley) and casein (dairy), autistic children showed marked decreases in autistic behaviors and improvements in communication and social skills with a reappearance of autistic behaviors when the diet was broken. Research suggested that these children had high levels of endogenous opioids in their systems due to ineffectual digestion of foods containing gluten and casein, a condition attributed to opioid excess (ibid).

Beaubrun and Gray (2000) reviewed herbs commonly used for psychiatric symptoms including St.John's wort, kava, ginkgo biloba, and valerian. Although evidence of the efficacy of herbal preparations in treating psychiatric conditions is growing, translating the results of efficacy studies into effective treatments for patients has been hampered by the chemical complexity of the products, the lack of standardization of commonly available preparations, and the paucity of well-controlled studies.

In addition, one study found that 15% of psychiatric patients were taking herbal medicines to treat their symptoms but none of them had communicated this information to their doctors, creating a dangerous potential for drug-herb interactions (Matthews, Camancho, Lawson, and Dimsdale, 2003). Like other treatments, natural remedies can have positive and negative effects, including a potential for serious drug interactions with traditional medicines. Therefore it's imperative that patients and physicians should be aware of these dangers.

32
Alternatives to Conventional Treatments
David's Story

David was a normally developing child until 15 months. At that time, he began to engage in repetitive behaviors and to withdraw socially. By 18 months, he had lost the ability to communicate and refused to make eye contact with other people. He ignored events happening around him and lived quite happily in his own world. After a series of evaluations, it was determined that he has autism, a disorder characterized by social and communication deficits as well as stereotypical behaviors. Autism is rising at such a rapid rate that it is sometimes referred to as the autism epidemic. According to the Centers for Disease Control (CDC), autism affects 1 in 167 children (CDC, 2004). Most etiologies of autism remain unknown, and the principal treatment is behavior modification. However, there is a growing body of empirical and anecdotal evidence to suggest that there are natural remedies for symptoms of the disorder as well.

The first of these treatments used for David was diet therapy. David was placed on a casein and gluten free diet that is used to address the opioid excess theory of autism and is becoming a first line defense in the treatment of autism. Like most autistic children, David had self-limited his diet to only foods containing gluten and casein and David exhibited the typical withdrawal reaction to the diet that is similar to withdrawal from other opioids including irritability, sleeplessness, and diarrhea. However, after six weeks, these symptoms disappeared and he began slow but steady improvements.

Eventually it was determined that casein did not have adverse effects on David's behavior and was reintroduced into his diet. He remains on the gluten free diet and shows marked regression when he accidentally ingests gluten. In a fascinating demonstration of the opioid effect of these foods, when David had a hernia operation, his mother gave him foods containing gluten in lieu of pain medications. David went from barely being able to walk following the surgery to climbing on the furniture following ingestion of the foods. The foods were discontinued a few days after the surgery.

David was also placed on an anti-yeast regimen involving dietary restrictions, antifungal medications, and the use of probiotic supplements for a short amount of time to eliminate candida yeast from his system. Like many autistic children, David had experienced chronic ear infections and had been on antibiotics for most of the first 15 months of his life. Introduction of echinacea drops finally eliminated these infections. Again he presented with a regression of symptoms due to the toxins released during the die-off of the yeast. However, once he finished the treatment, he no longer experienced bursts of hyperactivity that had occurred immediately following meals. David continues to take probiotic supplements to prevent yeast growth as well as a number of vitamin and mineral supplements specifically designed for autistic children.

David's mother used pet therapy to help David develop social skills. His love of cats helped him to develop empathy, and one of his regular household chores involved feeding his own cats. When taking his pets to the veterinarian's office, David routinely offers comfort to his cats and shows empathy by saying, "Ow, ow, ow" when they receive shots. Additionally, David's speech therapist reported bursts of language each time a new pet has been added to the household.

David continues to receive behavioral therapies through his local school system and attends elementary school alongside his peers with the help of an aide. He is able to communicate his needs and desires verbally and to follow simple instructions. He remains on the gluten free diet. He shows a remarkable amount of empathy for both animals and humans and engages in imaginative play.

References Part VI

Altschuler L (1999). Pet-facilitated therapy for posttraumatic stress disorder. Annals of Clinical Psychiatry 11(1):29-30.

Antoni MH (2003). Stress management effects on psychological, endocrinological, and immune functioning in men with HIV infection: empirical support for a psychoneuroimmunological model. Stress 6(3): 173-188.

Asheim Hansen B and Brodtkorb E (2003). Partial epilepsy with "ecstatic" seizures. Epilepsy and Behavior 4(6):667-73.

Baetz M, Larson DB, Marcoux G, Bowen R, and Griffin R (2002). Canadian psychiatric inpatient religious commitment: an association with mental health. Canadian Journal of Psychiatry 47(2):159-66.

Bakre MS and Morgan WP (1978) Anxiety reduction following exercise and meditation. Cognitive Therapy and Research 2:323-333.

Barlow J and Hainsworth J (2001). Volunteerism among older people with arthritis. Ageing and Society 21(2):203-217.

Barnes PM, Powell-Griner E, McFann K and Nahin RL (2004). Complementary and alternative medicine use among adults: United States, 2002. Advanced Data 27(343):1-19.

Baum A and Posluszn, DM (1999). Health psychology: mapping biobehavioral contributions to health and illness. Annual Review of Psychology 50:137-163.

Beachy JM (2003). Premature infant massage in the NICU. Neonatal Network 22(3):39-45.

Beaubrun G and Gray GE (2000). A review of herbal medicines for psychiatric disorders. Psychiatric Services 51(9):1130-1134.

Benjamins MR (2004). Religion and functional health among the elderly: is there a relationship and is it constant? Journal of Aging and Health 16(3):355-74.

Biddle S (2001) Exercise and depression. British Medical Journal. Available at: http://bmj.bmjjournals.com/cgi/eletters/322/7289/763 13653.

Blalock AC, McDaniel JS, and Farber EW (2002). Effect of employment on quality of life and psychological functioning in patients with HIV/AIDS. Psychosomatics: Journal of Consultation Liaison Psychiatry 43(5):400-404.

Blanchard CM, Stein KD, Baker F, Dent MF, Denniston MM, Courneya KS, and Nehl E (2004). Psychology and Health 19(1):1-13.

Boardman J, Grove B, Perkins R, and Shepherd G (2003). Work and employment for people with psychiatric disabilities. British Journal of Psychiatry 182(6):467-468.

Boey KW (2003). Religiosity and psychological well-being of older women in Hong Kong. International Journal of Psychiatric Nursing 8(2):921-35.

Bonadonna J Ramita (2002). Therapeutic touch. Handbook of Complementary and Alternative Therapies in Mental Health. Shannon S, ed, pp 231-248. San Diego, CA: Academic Press.

Borg J, Andree B, Soderstrom H, and Farde L (2003). The serotonin system and spiritual experiences. American Journal of Psychiatry 160(11):1965-9.

Boudreaux ED, O'Hea E, and Chasuk R (2002). Spiritual role in healing. An alternative way of thinking. Primary Care 29(2): 439-454.

Brodie SJ and Biley FC (1999). An exploration of the potential benefits of pet-facilitated therapy. Journal of Clinical Nursing 8:329-337.

Brown WS (2002). Psychoneuroimmunology and Western religious traditions. In In Link between religion and health: Psychoneuroimmunology and the faith factor. Koenig HG and Cohen HJ, eds, pp.262-274. London: Oxford University Press.

Callaghan P (2004). Exercise: a neglected intervention in mental health care? Journal of Psychiatric Mental Health Nursing 11(4):476-83.

Centers for Disease Control (2004). http://www.cdc.gov.

Chang BH, Skinner KM, Zhou C, and Kazis LE (2003). The relationship between sexual assault, religiosity, and mental health among male veterans. International Journal of Psychiatry 33(3):223-39.

Cohen S (2002). Psychosocial stress, social networks, and susceptibility to infection. In Link between religion and health: Psychoneuroimmunology and the faith factor. Koenig HG and Cohen HJ, Eds, pp.101-123. London: Oxford University Press.

Coleman CL (2004). The contribution of religious and existential well-being to depression among African American heterosexuals with HIV infection. Issues in Mental Health Nursing 25(1):103-10.

Comings DE, Gonzales N, Saucier G, Johnson JP, and MacMurray JP (2000). The DRD4 gene and the spiritual transcendance scale of the character temperament index. Psychiatry and Genetics 10(44):185-9.

Cramer EH, Jones P, Keenan NL, and Thompson BL (2003). Is naturopathy as effective as conventional therapy for treatment of menopausal symptoms? Journal of Alternative Complementary Medicine 9(4):529-538.

Dickerson SS, Kemeny, ME, Aziz N, Kim KH, and Fahey JL (2004). Immunological effects of induced shame and guilt. Psychosomatic Medicine 66(1):124-131.

Diego MA, Field T, Sanders C, and Hernandez-Reif M (2004). Massage therapy of moderate and light pressure and vibrator effects on EEG and heart rate. International Journal of Neuroscience 14(1):31-44.

Dieter JN, Field T, Hernandez-Reif M, Emory EK, and Redzepi M (2003). Stable preterm infants gain more weight and sleep less after five days of massage therapy. Journal of Pediatric Psychology 28(6): 403-411.

DiLorenzo TM, Bargman EP, Stucky-Ropp R, Brassington GS, Frensch PA and LaFontaine T (1999) Long-term effects of aerobic exercise on psychological outcomes. Preventive Medicine 28:75-85.

Ensel W and Lin N (2004). Physical fitness and the stress process. Journal of Community Psychology 32(1):81-101.

Esch T, Guarna M, Bianchi E, Zhu W., and Stefano G B (2004). Commonalities in the central nervous system's involvement with complementary medical therapies: Limbic morphinergic processes. Medical Science Monitor 10(6):MS6-MS17.

Etnier J, Salazar W, Landers DM, Petruzzello SJ, Myungwoo H and Nowell P (1997). The influence of physical fitness and exercise upon cognitive functioning: a meta analysis. Journal of Sport and Exercise Psychology 19:249-277.

Faver CA (2004). Relational spirituality and social caregiving. Social Work 49(2):241-9.

Fitzpatrick TR and Bosse R (2000). Employment and health among older bereaved men in the normative aging study: One year and three years following a bereavement event. Social Work in Health Care 32(2):41-60.

Fried KP (1996). Pet-facilitated therapy as adjunctive care for home hospice patients: A human service program design to promote quality of life. Dissertation Abstracts International: Section B: The Sciences and Engineering 57(5-B):3409.

Goldsmith AH, Veum JR, and Darity W (1997). Unemployment, joblessness, psychological well-being and self-esteem: Theory and evidence. Journal of Socio-Economics 26(2):133-158.

Griffiths PJ (2002). Psychoneuroimmunology and Eastern religious traditions. Link between religion and health: Psychoneuroimmunology and the faith factor. In Link between religion and health: Psychoneuroimmunology and the faith factor. Koenig HG and Cohen HJ, eds, pp.250-261. London: Oxford University Press.

Grossman P, Niemann L, Schmidt S, and Walach H (2004). Mindfulness-based stress reduction and health benefits: a meta-analysis. Journal of Psychosomatic Research 57(1):35-43.

Haas HS and Schauenstein K (1997). Neurimmunomodulation via limbic structures-the neuroanatomy of psychoimmunology. Progressive Neurobiology 51(2):195-222.

Hooker SD, Holbrook Freeman L, and Stewart P (2002). Pet therapy research: A historical review. Holistic Nursing Practices 16(5):17-23.

Jackson AP, Brooks-Gunn J, and Blake M (1998). Employment status, psychological well-being, social support, and physical discipline practices of single Black mothers. Journal of Marriage and the Family 60(4):894-902.

Kaminski M, Pellino T, and Wish J (2002). Play and pets: The physical and emotional impact of child-life and pet therapy on hospitalized children. Children's Health Care 31(4):321-335.

Kehoe M. (1990). Loneliness and the aging homosexual: Is pet therapy an answer? Journal of Homosexuality 20(3-4):137-142.

Khilnani S, Field T, Hernandez-Reif M, and Schanberg S (2003). Massage therapy improves mood and behavior of students with attention-deficit hyperactivity disorder. Adolescence 38(152):623-638.

Kiecolt-Glaser JK (1998). Psychological influences on surgical recovery: perspectives from psychneuroimmunology. American Psychologist 53(11):1209-1218.

Kiecolt-Glaser JK (2002). Emotions, morbidity, and mortality: New perspectives from psychneuroimmunology. Annual Review of Psychology 53(1):83-107.

Kiecolt-Glaser JK, Robles TF, and Glaser R (2002). Psychoneuroimmunology: psychological influences on the immune function and health. Journal of Consulting and Clinical Psychology 70(3):537-547.

Kim S and Feldman DC (2000). Working in retirement: The antecedents of bridge employment and its consequences for quality of life in retirement. Academy of Management Journal 43(6):1195-1210.

King DE and Bushwick B (1994). Beliefs and attitudes of hospital inpatients about faith healing and prayer. Journal of Family Practice 39(4):349-352.

Knivsberg AM, Reichelt KL, and Nodland M (2001). Reports on dietary intervention in autistic disorders. Nutrition Neuroscience 4(1): 25-37.

Knivsberg, AM, Reichelt KL, Hoien T, and Nodland M (2002). A randomized, controlled study of dietary intervention in autistic syndromes. Nutrition Neuroscience 5(4):251-261.

Koenig HG, George LK, Titus P, and Meador KG (2004). Religion, spirituality, and acute care hospitalization and long-term care use by older patients. Archives of Internal Medicine 164(14):1579-85.

Krau A and Walld R (2003). Influence of lack of full-time employment on attempted suicide in Manitoba, Canada. Scandinavian Journal of Work, Environment, and Health 29(1):15-21.

Lawlor DA and Hopker SW (2001) The effectiveness of exercise as an intervention in the management of depression: systematic review and meta-regression analysis of randomised controlled trials. British Medical Journal 322:763-766.

Levin J (2002). Is depressed affect a function of one's relationship with God? Findings from a study of primary care patients. International Journal of Psychiatry and Medicine 32(4):379-93.

Mabe PA and Joseophson AM (2004). Child and adolescent psychopathology: spiritual and religious perspectives. Child and Adolescent Psychiatric Clinics of North America 13(1):111-25, vii-viii.

Mansfield CJ, Mitchell J, and King DE (2002). The doctor as God's mechanic? Beliefs in the Southeastern United States. Social Science Medicine 54(3):399-409.

McAuley E, Jerome GJ, Marquez DX, Elavsky S, and Blissmer B (2003). Exercise self-efficacy in older adults: social, affective, and behavioral influences. Annals of Behavioral Medicine 25(1):1-7.

McCurdy EA, Spangler JG, Wofford MM, Chauvenet AR, and McLean TW (2003). Religiosity is associated with the use of complementary medical therapies by pediatric oncology patients. Journal of Pediatric Hematological Oncology 25(2):125-129.

McVarish CA (1995). The effects of pet-facilitated therapy on depressed institutionalized inpatients. Dissertation Abstracts International: Section B: The Sciences and Engineering 55(7-B):3019.

Means KM, O'Sullivan PS, and Rodell DE (2003). Psychosocial effects of an exercise program in older persons who fall. Journal of Rehabilitation Research and Development 40(1):49-58.

Merino Navarro D, Garcia Melchor M, Palomar Gallardo C, and Cano Lopez MC (2002). Massages and premature child care. Rev Enferm 25(6):12-4.

Milstein G (2004). The link between religion and health: Psychneuroimmunology and the faith factor. American Journal of Geriatric Psychiatry 12(3):332-334.

Moneymaker JM and Strimple EO (1991). Animals and inmates: a sharing companionship behind bars. Journal of Offender Rehabilitation 16(3-4):133-152.

Morgan WP and O'Connor PJ (1988) Exercise and mental health. In Exercise Adherence: Its Impact on Public Health, Dishman RK, ed, pp 91-121. Human Kinetic Books, Champaign, IL.

Morrow-Howell N, Hinterlong J, Rozario PA, and Tang F (2003). Effects of volunteering on the well-being of older adults. Journals of Gerontology Series B: Psychological Sciences and Social Sciences 58(3):S137-S145.

Moyer CA, Rounds J, and Hannum JW (2004). A meta-analysis of massage therapy research. Psychology Bulletin 130(1):3-18.

Mueller PS, Plevak DJ and Rummans TA (2001). Religious involvement, spirituality, and medicine: implications for clinical practice. Mayo Clinic Proceedings 76(12):1225-35.

Muschel IJ (1984). Pet therapy, with terminal cancer patients. Social Casework 65(8):451-458.

Musick MA and Wilson J (2003). Volunteering and depression: the role of psychological and social resources in different age groups. Social Science Medicine 56(2):259-269.

Mutrie N (2002) Healthy body, healthy mind? Psychologist 15:412-413.

Oman RF and Duncan TE (1995). Women and exercise: An investigation of the roles of social support, self-efficacy, and hardiness. Medicine, Exercise, Nutrition, and Health 4(5):306-315.

Parcel TL and Menaghan EG (1997). Effects of low-wage employment on family well-being. Future of Children 7(1):116-121.

Pert CB and Bowie DL (1979) Behavioural manipulation of rats causes alterations in opiate receptor occupancy. In: Endorphins in Mental Health, Usdin E, Bunney WE, and Kline NS, eds, pp. 93-104. Oxford University Press, New York.

Petruzzello SJ, Landers DM, Hatfield BD, Kubitz KA and Salazar W (1991) A meta-analysis on the anxiety-reducing effects acute and chronic exercise. Sports Medicine 11:143-182.

Post-White J, Kinney ME, Savik K, Gau JB, Wilcox C, and Lerner, I (2003). Therapeutic massage and healing touch improve symptoms in cancer. Integral Cancer Therapy 2(4):332-344.

Rabin BS (2002). Understanding how stress affects the physical body. Link between religion and health: Psychoneuroimmunology and the faith factor. In In Link between religion and health: Psychoneuroimmunology and the faith factor. Koenig HG and Cohen HJ, Eds, pp.43-68. London: Oxford University Press.

Reimer DF (1999). Pet-facilitated therapy: An initial exploration of the thinking and theory behind innovative intervention for children in psychotherapy. Dissertation Abstracts International: Section B: The Sciences and Engineering 60(5):2363.

Rojas R, Schlicht W, and Hautzinger M. (2003). Effects of exercise training on quality of life, psychological well-being, immune status, and cardiopulmonary fitness in an HIV-1 positive population. Journal of Sport and Exercise Psychology 25(4):440-455.

Rolniak S, Browning L, Macleod BA and Cockley P (2004).Complementary and alternative medicine use among urban ED patients: prevalence and patterns. Journal of Emergency Nursing 30(4):318-24.

Ross SB (1983). The therapeutic use of animals with the handicapped. International Child Welfare Review 56:26-39.

Sarrell EM, Cohen HA, and Kahan E (2003). Naturopathic treatment for ear pain in children. Pediatrics 111(5):574-579.

Sathyaprabha TN, Murthy H, and Murthy BT (2001). Efficacy of naturopathy and yoga in bronchial asthma-a self controlled matched scientific study. Indian Journal of Physiological Pharmacology 45(1):86-86.

Schapman AM and Inderbitzen HM (2002). The role of religious behaviour in adolescent depressive and anxious symptomatology. Journal of Adolescence 25(6):631-43.

Schwartz C, Meisenhelder JB, Ma Y, and Reed G (2003). Altruistic social interest behaviors are associated with better mental health. Psychosomatic Medicine 65(5):778-85.

Speigel D and Fawzy FI (2002). Psychosocial interventions and prognosis in cancer. Link between religion and health: In Psychoneuroimmunology and the faith factor. Koenig, HG and Cohen HJ, eds pp.84-100. London: Oxford University Press.

Steyer TE, Freed GL, and Lantz PM (2002). Medicaid reimbursement for alternative therapies. Alternative Therapies in Health Medicine 8(6):84-88.

Sullivan-Blum CR (2004). Balancing acts: drag queens, gender and faith. Journal of Homosexuality 46(3-4):195-209.

Van der Zouwe N, van Dam FS, Aaronson NK, and Hanewald GJ (1994). Alternative treatments in cancer; extent and background of utilization. Ned Tijdschr Geneeskd 138(6):300-306.

Van Sickle TD, Hersen M, Simco ER, Melton MA and Van Hasselt VB (1996) Effects of physical exercise on cognitive functioning in the elderly. International Journal of Rehabilitation and Health 2:67-100.

Vanden Boom DC and Lustig DC (1997). The relationship between employment status and quality of life for individuals with severe and persistent mental illness. Journal of Applied Rehabilitation Counseling 28(4):4-8.

Walach H, Guthlin C, and Konig M (2003). Efficacy of massage therapy in chronic pain: a pragmatic randomized trial. Alternative Complementary Medicine 9(6):837-846.

Weze C and Leathard HL (2004). Evaluation of healing by gentle touch for the treatment of musculoskeletal disorders. American Journal of Public Health 94(1):50-52.

Williams RB. (2002). Hostility, neuroendocrine changes, and health outcomes. Link between religion and health: Psychoneuroimmunology and the faith factor. In Link between religion and health: Psychoneuroimmunology and the faith factor. Koenig HG and Cohen HJ, eds, pp.160-173 London: Oxford University Press.

Zemore SE and Kaskutas LA (2004). Helping, spirituality, and Alcoholics Anonymous in recovery. Journal of Studies in Alcohol 65(3):383-91.

End Note

This book has introduced a vast range of material. As in many endeavors of its kind, it has presented a large number of topics but "covered" none. Every area of inquiry pursued by the researchers whose work is mentioned here has been studied in depth by at least a few scholars, and for several of the topics, by a large number. I found so many threads that would be fascinating to follow—but that's for future adventures. Yes, doing the research for this book has been an amazing adventure for me. I hope this introduction to neuroscience has also been an exciting journey for you.

INDEX